MW01005026

Rethinking Intelligence

ALSO BY DR. RINA BLISS

Race Decoded

Social by Nature

Rethinking Intelligence

A Radical New Understanding *of* Our Human Potential

DR. RINA BLISS

HARPER WAVE
An Imprint of HarperCollins*Publishers*

 For information, address HarperCollins Publishers, 195 Broadway, New York, NY 10007.

HarperCollins books may be purchased for educational, business, or sales promotional use. For information, please email the Special Markets Department at SPsales@harpercollins.com.

FIRST EDITION

Designed by Nancy Singer

Library of Congress Cataloging-in-Publication Data has been applied for.

ISBN 978-0-06-323778-0

23 24 25 26 27 LBC 5 4 3 2 1

For Nene

Contents

Rethinking Intelligence

Introduction

Being intelligent was my mantra. I believed it was my ticket out.

I grew up in a middle-class home in a middle-of-the-road neighborhood in Los Angeles's San Fernando Valley. Our public schools provided a mediocre education to an average sample of LA youth (which was remarkably diverse, thanks to busing laws that helped to redistribute the city's population). At first glance, there was nothing noteworthy nor problematic about my situation. But prick the surface and a different image appears.

My family was struggling. My mom, Liza (née Guojia Hua Caryabudi), hailed from the floating city of Banjarmasin in Indonesia. Distant relatives had snatched her from her young laboring parents and planted her in a Dutch convent on the neighboring island of Java. In college, she joined a generation of Western-enculturated Indonesians who shipped off to Europe and the United States to study abroad. When her scholarship ran out, she took up work as a domestic servant in the Hollywood Hills. She vowed to do anything to remain.

My dad, Nathaniel Jr. (also known as "Junior" or "Natty"),

belonged to one of the well-to-do families that lived in those Hills, a prominent military family in California politics. But beneath a veneer of Democratic respectability lay a family ravaged by trauma, abuse, and addiction. Tormented by prisoner-of-war flashbacks, my grandfather terrorized my father with knockout blows and fugues of rage. By eight, my dad was regularly robbing his parents' medicine cabinet of sleeping pills and painkillers. By eighteen, he was subsisting on a daily diet of codeine and barbiturates.

My mom and dad met one balmy day in 1972, as they were descending the hills for the valley below. They flirted at the bus stop at Hollywood and Vine and then boldly arranged to meet again. It was love at first sight.

Lifted by love, my parents tried to make a fresh start. My mom took a secretarial course and my dad started law school. They rented a sunny apartment in Sherman Oaks. Yet by the time I came around, substance abuse had overtaken my dad, and it had moved him to the streets, to the slippery recesses of junk dens and crack houses. As he spiraled, my mom ferried us to a safer place. She got an office job in Beverly Hills, rented a new apartment just for the two of us, and found someone to watch me day and night as she hustled overtime to support us.

It was hard waking up to my mom leaving every morning and heartbreaking to fall asleep before she got home. My dad would come over to play with me and the other little ones under the care of our building's sitter, but he was always on the brink of an overdose and often unintelligible.

Shockingly, however, my real troubles began when I started school. As the only Southeast Asian biracial kid in my classes, I was immediately regarded as an oddball—not white, not black, not even yellow . . . an indeterminate tawny "other." One boy in my kindergarten class made a game of rounding up a group of kids to chase me around the playground, making squint-eyed faces and hurling epithets at me. Another kid at my summer program would cut me off when I spoke, to deliver slurs and threats. Negative stereotypes abounded and crowded the air I breathed every day.

The one positive about those stereotypes: kids thought I was *smart*. Being Asian (even just part) seemed to magically confer upon me a superior intellect. As one school year gave way to another, I took solace in my perceived giftedness. Believing there was no other way to go, I read, I studied, I achieved, and I outsmarted.

Or so I thought.

The way I saw intelligence as a kid was limited, to say the least. I believed being smart was an innate quality, and this is exactly how grown-ups around me talked about it. In the classroom, I heard the word *gifted* just about every time anyone talked about intelligence. My teachers and TAs encouraged only some of us students to dream of being intellectuals and growing up to use our smarts. Others they would applaud for their athletic ability, as if being intelligent and physically adept were opposing things. In the media, scientists were intelligent. Doctors were intelligent. Professors were intelligent. Engineers, too. In my eyes, these high achievers were winning at life as a result of their God-given talents, with little exertion required from them.

My understanding of it was you were either born with intelligence, or not. It certainly wasn't something that anyone could learn and develop as a skill over time.

Growing up with American media, the stereotypes I saw in TV and movies certainly didn't help with this misconception: between the "autistic-coded math genius" and the "East Asian tech wiz," brainy, intelligent characters were either white men (and the main character) or people of color (in a supporting role).

Popular science only made matters worse. For centuries, evolutionary biology, genetics, phrenology, and psychology popularized the idea that, like height and weight distribution, intelligence ran in families. It was etched into our DNA, a part of our innate biology. Scientists had worked hard to perfect intelligence measurement. A simple IQ test said it all and foretold our future.

Placing Intelligence

In the 1980s and '90s, intelligence testing—based on a universal criterion of "intelligence quotient"—was big business, and its implementation was ubiquitous. School districts like mine were testing all year long to determine teacher performance, student performance, and student giftedness. High test scores meant more resources—better funding, better teachers, and more support for specialized programs. From administrators to parents, everyone was deeply invested in the outcome of these tests.

Even for a "smart kid" like me, the pressure to perform was overwhelming. I was tormented by the fear that I wouldn't measure up. And then one day, my fear came true.

A letter from the LA Unified School District arrived stating that I had not tested into the local magnet school. Instead, I would have to remain at my run-of-the-mill neighborhood elementary school. The news devastated my mother. It was the first time I ever saw her cry.

Soon after, she purchased two thin Mensa books, puzzle books designed by the world's oldest high-IQ society. She would quiz me from them in the evenings, sometimes late into the night. As much as I loved spending that time with her, I was confounded by the kinds of questions I found in those books. I was missing background information on the words or shapes, and I recognized very few of them. But I played along and tried my best, because I was terrified my ineptitude might make my mother cry again.

On test day, I squeezed my pudgy fist around the thick blue pencil. A sour breath rushed into my throat. Letters and shapes grew fuzzy as I struggled to make out the exam's questions. But I got lucky. Though I didn't test out of my school, I performed well enough to test out of the standard curriculum. By the third grade, I was one of the few students who were tutored apart from the rest of the class and cultivated for educational excellence. Slurs gave way to accolades. And though the racial stereotyping around me only intensified, I found that overachievement could also garner praise. It protected me, at least emotionally, from the sting of being different.

My classmates started to suggest that I was the smartest kid in the class. My teachers echoed them. I was proud to come home to my mom and pull the Outstanding Reader award from my backpack, my heart swelling as she smiled. Intelligence became my mantra. There was never any question of whether I would prioritize academics, or graduate, or go to college, or get a degree and a job that utilized my inherent smarts. Even when my family moved from LA to the Coachella Valley, and I began skipping school in a classic display of teenage rebellion, I still made it a point to show up for exams. I would get that A-plus and preserve my sterling GPA. I didn't question any of it—the ends, the means, the definition of intelligence at its heart. To me it all made perfect sense. I was using my smarts to propel me to a new place.

At the University of California, Santa Cruz, my perspective shifted. I chose from courses based on their competencies, not their subject. I took a film class to fulfill my writing requirement, an oceanography class to satisfy my quantitative reasoning, a philosophy class to cover human behavior. In place of grades, I received long narrative evaluations that explained my engagement with my courses in detail. I was freed from the rote academic scheme that had characterized my prior education.

I began to think critically about curriculum. I began seeing it as information worth questioning. I began to think reflexively, looking at my world, and my position in it, with doubt. In other words, I began thinking like a scientist.

I also started to probe the status quo on race. Knowing full well that others' views of me never matched up with how I viewed myself, or my real cultural and ancestral heritage, I began doubting the biological "truth" of so-called "races." Where did the idea of race even come from? Was my race really just a function of my DNA?

And if my race wasn't what I thought it was, then what about my intelligence? I still believed that I was inherently smart. But I also started to wonder what it meant to be "smart" as I befriended students who had worked even harder than me to get through high school and into college. I learned of classrooms so underfunded and inadequate that their students needed tutoring all-nighters to make up for it. I heard about how hard things got when jobs quashed a student's chance to study after school. On top of this, some of my friends felt unsafe where they went to school, in their neighborhoods, and at work. And for my Black and Indigenous friends, their race didn't confer an innate sense of superiority; most of them were never told that they were gifted. In fact, many of them were told the opposite.

I devoted the next decade of my life to researching the science and history of race. Over that time, I learned that humans had invented the notion of race long before we ever had a theory of genes, way before science or the field of biology even existed. The first mention of race appeared around the emergence of global travel and mercantile capitalism in the fifteenth century, when European fleets began sailing around the world and encountering what to them were new peoples. Explorers, conquerors, and enslavers brought back stories of these other cultures. Then self-fashioned naturalists and armchair zoologists used those tales to conjure up human taxonomies. What they referred to as "nature's races."

Races were characterized as original human peoples who, at some point in time, had lived and bred apart from one another on different continents. Races had supposedly evolved their own defense mechanisms for their unique environments, as well as advantageous behavioral adaptations that helped them survive those environments. But there was never any testing of these assumptions. Ruling powers simply adopted racial classifications even in the absence of scientific proof.

I was even more taken with the biology of race that emerged in the eighteenth century as genes were being discovered. I read everything from Charles Darwin's *On the Origin of Species* to Spencer Wells's *The Journey of Man*. I took a private crash course in human genetics in which I trained with a geneticist who would become one of Columbia University's premier stem cell neuroscientists, and I completed a systematic analysis of contemporary genetic science reports on human variation.

Pipetting cells into a centrifuge in the lab, and decoding patent applications on novel DNA software programs, I felt the electrifying thrill of discovery. It was now clear to me that my doubts about race were confirmed. I wasn't born to be one way or another, and I wasn't just a stereotype. None of us were.

My first book, *Race Decoded,* summarized my main finding, that genetic populations do not correspond to the invented taxonomies we call "human races." Instead, DNA analysis of people living all over the globe proved my suspicion to be true: Race is a social fiction that has only been made fact due to our false belief in it. It is an unscientific idea conjured up to create hierarchies of power. It is no stand-in for genetic ancestry.

Following the release of *Race Decoded,* I joined the faculty of the University of California, San Francisco, and began teaching a diverse group of young scientists, medical students, medical sociologists, and medical humanities scholars about our scientific blind spots on human differences. I also began working with leading geneticists and genome mappers to publicize my findings and educate people on the truth about genetic differences. I became a regular commentator on race and genetics, speaking with journalists, filmmakers, and correspondents all over the world.

But something was gnawing at me. It's true, my work had proved that our common belief that race was a good proxy for genetics was plain wrong. But there still was controversy around the specific characteristics associated with race. Geneticists were unearthing more and more variants associated with behavioral traits like educational attainment, aggression, delinquency, and intelligence. I jumped on this research and wrote my second book, *Social by Nature.*

I learned that a genetic science of social behavior, what I termed "sociogenomics," was emerging. While most sociogenomic studies were not comparing races, studies were making claims about our DNA that reinforced the notion that some of us were born with stronger, more functional genes while others were born with weaker, more dysfunctional ones. In the area of intelligence, researchers were trying to predict intellectual performance and life outcomes from our DNA. Some were arguing to arm the entire education system with DNA tests to track students with dysfunctional genes for success, as if our genetic code predestined us for success or failure.

Even though many scientists wanted to use DNA to help flag problems, get treatment, and identify weaknesses in policies and programs, we were still in the dark. There were too many genetic variants suspected to affect performance on intelligence tests. And we had no idea what the variants did in our brains and bodies. Without full knowledge of the genes at hand, and the interactions they have with people's unique environments, it was too early to talk about prediction and policy.

Nature Nurture Nescience

Then I got pregnant with twins. Identical twins. Two individuals who share a genome.

As my husband, Nick, and I watched the screen in the doctor's office, ultrasound wand skimming my gelled abdomen, two little peanut-shaped sprouts appeared. With gleeful tears spilling down our cheeks, we gave each other's hand a squeeze. We always knew we wanted children, but never in our wildest dreams did we think we would be welcoming two into the world at once!

I found myself staring at my work (and now at my growing belly) with a serious case of Beginner's Mind. It hit me hard just how little I, the genetics and society expert, knew about the humans taking form in my body. What did it mean that my babies shared a genome? How would they be similar? How would they be different? I was forced to rethink the balance of nature and nurture, genes and environments, and biology and society.

Knowing two were on the way, the complete set we had thought would come in sequence, this pregnancy felt like our *one big chance* to learn everything. (We didn't know then that we would be pregnant again a year later!) With the tidal wave of parenthood looming, we dove into the full spectrum of popular parenting literature, reading it with more grit and dedication than ever before.

As an expert in genetics, I was struck by the discrepancies between the science I had just written about versus the information in popular parenting and twin-specific manuals. The parenting literature told me that, counter to what intelligence researchers were saying, my boys' DNA would not be their destiny. Having the same genome would not make them behave in the same way. Their environments would be different enough to produce different personalities, different tendencies. Their environments would even shape their intellectual interests and academic pursuits.

In place of genetic testing, the parenting literature encouraged nurture as a way to nourish the brain. Parenting had become all about neuro-optimization, pledging better outcomes from a deeper knowledge of their child's brain architecture. Whether recommending a sling to tether my little one tight to my heart or keeping them at arm's length in a stroller, child-rearing gurus promised that such choices would lead my twins to develop the neural qualities that would ensure their ability to reach their potential.

But just like intelligence genetics, the parenting literature was racked with faulty assumptions about genes, neurobiology, and heredity. While brain-boosting specifically for intelligence was discussed less frequently in these resources, a child's overall success and well-being, along with "smart" and "intelligent," guarantees were everywhere. Higher IQs, improved memory, enhanced executive control, verbal and quantitative competencies, and even academic success were within reach for your child, if you followed the experts' advice. And in the long term, I was assured that this kind of brain sculpting would materialize self-driven successful humans, as if high IQ equaled surefire success.

More troubling than these IQ promises was that behind so many big headlines on child development and intelligence lay some kind of bunk study on twins. So, while I stocked up on "brain-boosting" DHA fish oil, and while Nick played "neuro-supportive" gamelan to my belly, I was critically analyzing the evidentiary basis of all the studies upon which this guidance was based.

Sitting with the literature, I was reminded of my early college days, when I began questioning the systems and institutions that shape our attitudes and beliefs. Being smart wasn't just about acing tests or having a high IQ. Knowing things wasn't just about acquiring and storing information. Aptitude wasn't just about the architecture of the brain. Intelligence was actually about adaptability, creativity, and resilience.

The way we were talking about intelligence, us grown-ups, and especially us experts in genetics and brain development, left me with a sinking sense that we had not evolved our thinking much, if at all. Though the neuro-optimization in the baby books I read looked different from my mom's Boomer-generation Mensa pursuits, the pressure to sculpt inborn smarts as a means of boosting IQ was the same. What's worse, as parents we were being encouraged to shape neurological development in utero, even though we had so little information about the relationship between intelligence and genes or genes and their environments.

This is when I realized it was time for a new approach to intelligence.

Rethinking Intelligence

This book is the result of more than a decade of research, as I worked toward a better understanding of the nature of intelligence. I found an innovative new area of genome science that has been making bold strides in our knowledge of intelligence and our understanding of the role our environment plays in shaping who we are. As I've pored over the latest genetic science, I've been able to realign my passion for intelligence with my passion for critical thinking.

I have learned that intelligence isn't about inherent mental superiority (or inferiority). And it's not a matter of achieving some genetic destiny. It's certainly not something that you can measure with a simple test or by sequencing your genetic code. Intelligence is about being aware of the world around you, learning from your environment, and communicating with others to conceive of new and better ways forward. It's about perceiving new knowledge opportunities and waking up to your world. It's about being curious and following a desire to explore the minutiae of human life.

Intelligence is a process. It is *in* process. It's a journey defined by change. It is always growing as you develop greater awareness of your world and shift your perceptions to utilize your knowledge. Most importantly: No one can score you on it. And nothing can take it away from you.

PART I

Understanding Intelligence

CHAPTER 1

Thinking Intelligence

in · tel · li · gence
/in'teləjəns/

Intelligence. Trait? Attribute? Condition? State of play? It's something quintessential to your being, entrenched in your everyday existence and who you are. And yet, like other immaterial matters, it's an "I know it when I see it" kind of thing. Hard to define yet intrinsic to life itself.

A web search conjures up an array of definitions: "the ability to acquire knowledge and apply knowledge and skills," or "the capacity for learning, reasoning, understanding, and similar forms of mental activity," or "a mental quality that consists of the abilities to learn from experience, adapt to new situations, understand and handle abstract concepts, and use knowledge to manipulate one's environment." Some definitions emphasize emotional know-how, adaptability, curiosity, even spirituality. Others focus on mental machinations and cognitive function.

While definitions differ, people tend to agree that

intelligence is about being able to use your mind to engage with your surroundings. It's about ability, capacity, aptitude, and talent. Intelligence is something concrete you have, inside your mind, at your disposal, no matter where you happen to be.

Yet a closer look at our most-held assumptions about intelligence and their roots in patriarchal and white supremacist thinking will show just how little we know.

Popular Thinking on Intelligence

Though there are people who have dedicated their lives to researching the nature of intelligence, it doesn't take a specialized level of understanding to comprehend the basics. Common understanding has it that intelligence is what sets us apart from other organisms. While many animals are social, and various species have diverse languages and technologies, only humans can form sentences or invent a computer program. While early hominids wore clothes, used tools, and lived in societies like we do, only *Homo sapiens* compose music and create works of art.[1] Humans are born with the capacity to learn, know, grow, and advance culture in ways far beyond what other animals can do, and we have evolved to use our knowledge to make our environment work for us.[2]

It is also common understanding that, despite being intrinsic to human nature, intelligence varies by individual. As a result, intelligence is something we work hard to quantify. When *Business Insider* releases the Smartest People of All

Time or *Forbes* issues the Smartest Countries in the World, they consult the numbers on IQ scores, math and science test scores, number of Nobel Prizes, higher education degrees, and other conventional measures of intellect. Intelligence quotient (IQ) score, specifically, is the predominant metric used to measure intelligence, and it is seen as a standalone definitive marker of aptitude. Academic performance and educational attainment rates also contribute to ranking. Centuries of testing individuals, averaging scores, and comparing them show that a small percentage of people around the world score high or low on intelligence indices, while the vast majority rank closer to the mean.

All around the world, we venerate those considered to be of high-ranking intelligence. Masterminds from the past, Einstein, Curie, as well as those from the present, Hawking, DeGrasse Tyson, Jobs, are more than mere household names. They are our heroes. They display a unique capacity to handle complex theories, sequences of information, and equations, and leverage their knowledge to be exceptionally productive. They use their intelligence to change the world.

Our fascination with intelligence is especially acute in terms of unsung genius, where we enjoy the public reveal of "girl next door" prodigies and "man on the street" mavens. Cheering on a *Shark Tank* feeding frenzy as a DIYer reveals their latest tech innovation, feeling the rush of watching preteens dominate at the National Spelling Bee. We love to witness supersmart people as they think on their feet, because we understand that this is human activity at its best.

But of course, intelligence is far more to us than public fascination. There's a private, highly personal dimension to our interest in it. We want to understand intelligence so that we can make the most of it, so that we ourselves can benefit from it in our day-to-day lives, and pass that benefit to our children.

Research confirms that intelligence can be wind in our sails, propelling us to better circumstances. Higher intelligence has been associated with a higher income, higher education level, more prosperity, and a longer life.[3] It also materializes as an increased motivation to achieve, lead, make healthy choices, exercise creativity, and enjoy mental health. High IQ scores in particular have been correlated with a gamut of positive outcomes like artistic success,[4] educational attainment,[5] extracurricular achievement,[6] fitness,[7] occupational status,[8] and even marital success.[9] So, while we chuckle at Albert Einstein's hair or Steve Jobs's black turtleneck, we know that there is creativity and freedom in being intelligent, in being able to access and apply those mental powers to solve problems in our lives, and we know that good things tend to come from using it.

Even without keeping abreast of the latest in intelligence science, we also share a common understanding that intelligence is at its best when paired with ingenuity. As biographer Walter Isaacson has put it, "Smart people are a dime a dozen. What matters is creativity, the ability to apply imagination in almost any situation." Einstein was similarly famous for saying, "The true sign of intelligence is not knowledge but imagination." He believed that tapping into intelligence spurred curiosity, a willingness to identify your own limits, and a motivation to learn what you don't already know. Researchers validate these assertions, with studies documenting that highly intelligent people find it easier to think against the grain and challenge the status quo than less intelligent people.[10] Intelligence can inspire risk-taking with shrewdness, audacity with prudence, and a will to break through walls that confine the limits of our imagination. It can also lead an individual to live a better life for themself while building a better world for all humankind.

The Roots of Today's Thinking

Our contemporary thinking on intelligence springs from a long and rich lineage of inquiry into what makes us tick. Throughout human history, some of our greatest thinkers have strived to capture its essence. Like us, they cherished intelligence and placed it at the foundation of the story of what it means to be human. Also like us, they held out hope, against all odds, that each of us would make the most of what we had and use our intelligence for the good of all.

Yet we've also picked up some very bad ideas from early thinking on intelligence. We owe to ancient Greek philosophers such as Aristotle, Plato, and Socrates our habit of ranking intelligence among people, and seeing only some of us as truly intelligent.[11] Though these thinkers were responsible for establishing intelligence as a God-given right for all humans, they also introduced the notion that few humans possessed a *high degree* of intelligence.

Plato, in particular, argued that intelligence—something inseparable from one's strength of character and courage—was in such short supply that our social order formed something akin to a pyramid.[12] At the bottom were people of low intelligence, who were deficient in every other mental regard, and were suitable only for manual labor, such as farming or construction. In the middle were those with some intelligence, who might be capable of policing or defense work, where they would have some authority over and responsibility for the well-being of others. At the top were those blessed with high intelligence, who were meant to analyze, strategize, and govern. Only these lucky few were destined to lead.

It's also worth noting that Plato was the first to give us the idea of selective breeding of like-minded people, something echoed in today's market of genetic matchmaking. As Plato stated it, "The good must be paired with the good, and the bad with the bad, and the offspring of the one must be reared and of the other destroyed; in this way the flock will be preserved in prime condition."[13]

Roman thought has also colored our current way of thinking of intelligence, but in the opposite direction of inborn rigidity, toward optimization. Though like the ancient Greeks, Romans believed in the innate basis of intelligence, unlike their predecessors, they believed in the value of improving intellect with mental calisthenics. Similar to our approach to brain-boosting today, Romans encouraged those well-endowed with intelligence to use "mnemonic" devices, or mental images, that could shorthand more complex concepts.[14] Roman elites used such tools as a means to help strengthen their smarts, as well as for more practical pursuits, such as memorizing lengthy speeches.

From the philosopher-cum-naturalists of the Enlightenment Era, we have inherited some sinister variations on these themes, especially the habit of seeing intelligence as having a racial basis. Throughout the eighteenth century, from Uppsala to Konigsberg, scholars like the wigged and powdered Carl von Linné[15] and Immanuel Kant[16] lectured that intelligence was in greater supply among humans of a certain genetic ilk: Europeans. After all, at the time they were intellectualizing intelligence, explorers were sailing around the world colonizing lands and people. Sailors' tales of monsters and heathens and the gamut of human variation inspired these thinkers to categorize human beings into "races of man," and to rank those races by intelligence.

In addition to this view of the connection between race and intellect, we owe to seventeenth-century Enlightenment thinkers our belief that intelligence is a sign of genetic superiority, one worth optimizing if you happen to be the most genetically endowed.[17] They were the first to suggest—well before the advent of actual genetic science—that some enigmatic seed, an instruction manual deep in your body and soul, made you who you were, whether a rational European or an irrational (uneducable) Other. If you were the former, you had a responsibility to maximize your mind so that you could lead mankind to the promised land of civilization.[18] As the Age of Discovery sunsetted and the Industrial Revolution dawned, Late Enlightenment thought leaders like Georges Cuvier[19] and David Hume[20] said we must empower wellborn individuals to take over the world and subjugate the masses of unintelligent Others. They saw human progress as a world driven by slavery and genocide, where Europeans of high intelligence developed their culture and politics, and stamped out all other so-called "unreasonable" forms.

Though today we live in an era of human rights, where slavery and genocide are seen as crimes against humanity, it is no secret that racist views of intelligence continue to endure. We therefore remained tethered to these unscientific notions that would have us believe that high intelligence is the birthright of the few.

The Genius of Genes

Enter Darwin, to whom we owe our entire framework for life sciences. Journeying aboard the HMS *Beagle* in the years 1831–36, the father of genetics and modern biology spent weeks sailing through tropical waters, exploring archipelagos brimming with peculiar flora and fauna. In the glimmer of the South Pacific sun, he penned his theory of evolution by natural selection, that organisms fitter for a particular environment will survive to pass on their genes, thereby changing the nature of entire species.

Darwin also pondered the origin of intelligence, and

here his ideas were more evolutionary than revolutionary. Darwin echoed those thinkers who came before him, characterizing intelligence as the provenance of humankind but a quality that evolved to different degrees depending on one's race.[21] He surmised that humankind was a hierarchy of racial subspecies.

That hierarchy ran from, as he wrote, the "feebleminded African barbarian" to the rational European savant, or, in his view, "a savage who does not use any abstract terms, and a Newton or Shakespeare."[22] Indeed, Darwin was one of the first to introduce the concept of craniometry to human science, believing that our skulls measured differently based on our intellectual prowess. He argued that if a European happened upon a "Negro" skull, they would conclude it belonged to a different animal altogether.[23]

Darwin wasn't alone. He was surrounded by other prominent (white, male) scientists who cheered him on and awarded him accolades for his evolutionary theory, and it is from these men that we get many of our contemporary notions of mental fitness. In fact, it was the autodidact and polymath Herbert Spencer who coined the phrase "survival of the fittest," which is now common parlance.[24] Spencer built his entire career on what he famously called "Social Darwinism," or the belief that not only genes but classes, races, and cultures evolved through fitness, and that the fitter survived to dominate, subjugate, and eventually outlive the rest.

Darwin adopted Spencer's theory and called for controlled breeding of these fittest. He argued that allowing "the weak members of civilised societies [to] propagate their kind" would "be highly injurious to the race of man," leading to its "degeneration."[25] Those with low intelligence had to be prevented from passing on their genes. This could be achieved by enslavement, such as the slavery then found in the Americas, but also by abolishing protections for the sick and poor so that they could hasten to their ends.

Modern science quickly adopted the Darwinian belief that intelligence was genetically determined, and that belief has since infiltrated every scientific field, from biology and anthropology to sociology, politics, and economics. But it is Darwin's cousin, the petulant globetrotter Francis Galton, to whom we owe the mass popularization of this idea. Until his death in 1882, Darwin held out for the possibility of socializing "savages," thereby rendering them useful to the human race in some way, shape, or form. But the tightly wound Galton, who won every scientific accolade and was knighted by King Edward VII in 1909, believed so strongly that intelligence was absolute and unalterable that optimization could only happen through genetic purging, or, as he called it, "eugenics."[26]

More than Better

Eugenics was both a science and a social movement, as intensely personal as it was political. Its purpose was to encourage breeding among the "bright bulbs" and prevent the "dim" from doing so in order to promote the ideal populace.

This attempt to, as Galton put it, "produce a highly-gifted race of men" was an all-out public health campaign in Europe and the United States that materialized the desires of so many intellectuals and political elites who came before, as well as countless leaders who were in power at the time.[27] As a result of these efforts and inroads, eugenics became widespread and highly successful, implemented by governments around the world. Indeed, this method of intelligence-boosting dominated geopolitics for decades, well into the twentieth century.[28]

How exactly did eugenicists aim to increase intelligence? Domestically, they issued pro-natalist policies that incentivized members of the European upper classes to have children, and anti-natalist policies such as mass sterilization[29] and incarceration of those deemed mentally unfit.[30] Internationally, they argued for racial elimination programs via genocide and colonial settlement. Most know of the horrors of the Nazi concentration camps, wherein Nazis murdered upwards of 11 million Jews, Roma, political prisoners, LGBTQ and disabled people, and anyone else deemed unfit for society, but few know about America's eugenics program, which culminated in the forced sterilization of 65,000 so-called "feebleminded" citizens.[31] Fewer still know of Galton's call for a mass genocide of the entire African continent by moving in genetic warriors to breed Africans out of existence.[32]

Eugenics might seem like a leap, but as we discuss human intelligence, it is very much part of the conversation, especially when it comes to intelligence being regarded and measured in terms of personal scores. Indeed, it was originally a population science—a demographic science, even—aimed at encouraging brain-boosting at the species level.[33] But many of its tactics were individualized precisely because they were based in human sexuality and inspiring certain populations to take personal action to benefit their kind.

To this end, eugenicists introduced our love of intelligence testing. In the first two decades of the twentieth century, they organized publicly attended "Fitter Families"[34] and "Better Babies"[35] fairs where intelligent people could mingle, receive sex and marriage counseling, undergo intelligence testing (with many wanting to know and beat their high score), and compete to win prizes. Strolling through the sunny fairgrounds, one would encounter sanitation and nutrition booths offering instruction on optimal ways of living, stages outfitted with all kinds of tools and scales with which babies and adults could be measured and weighed, and offices where eyes, teeth, and all elements mind and body could be examined, rated, reported, and prescribed new regimens. Eugenicists called for the intellectually gifted to take matters into their own hands, to get tested, find similarly gifted mates, and boost their genetic stock.

Intelligence Gets a Retrofit

In the years following World War II, as the world struggled to comprehend the sheer scope and devastation of the

Holocaust, the eugenics movement attempted to rebrand itself from a science focused on stamping out low intelligence in the "lower races" to one that identified the potential for high intelligence across racial groups.[36] Few scientists, after all, wished to be associated with explicitly racist ideologies, let alone concepts central to genocide. Many worked hard to uncouple the notion of intelligence from race, and to salvage scientific definitions of intelligence and measurement tools by avoiding racial comparisons.

While racism and prejudice continued to permeate societies worldwide, a group of intelligence scientists set to redefine intelligence as malleable for all, and improvable with cultivation in the form of education, socialization, and behavioral change.[37] They argued that intellectual aptitude requires adaptability, flexibility, and open-mindedness. Anybody from any background can possess these qualities. In other words, intelligence was democratized.

In the fields of biology, education, and psychology, researchers began exploring how best to measure and quantify intelligence. Could it be captured by the score generated on IQ tests? Or was it more nuanced than that? At the twilight of the twentieth century, Harvard psychologist Howard Gardner offered his answer to this question by introducing a new theory of intelligence called "multiple intelligences."[38] This theory posits that there isn't just one, but rather many forms of intelligence. You might be excellent at math or writing, or you might struggle with those skills yet excel in others. Just think of those math savants who can't hold a conversation at a cocktail party. Or artistic virtuosos who struggle to calculate the tip after a meal.

Psychologist and science writer Daniel Goleman also offered a theory of multiple intelligences, and a pointed critique of IQ, with his bestselling book *Emotional Intelligence: Why It Can Matter More than IQ*. In this book, Goleman used the basics of brain anatomy to show how the parts of the brain that control our emotions can overtake our ability to reason and disempower our intellect (a phenomenon witnessable in high-IQ individuals who fail academically or those who perpetrate violent crimes). Goleman argued that instead of IQ testing, we need to focus on emotional testing, and emotional teaching and learning, to train our emotional minds to enable our rational minds to become more intelligent.

Though Gardner's and Goleman's theories have done little to dispel prior notions of heritability in intelligence science (as we will learn in the following chapter), they have had an irreversible effect on popular and scholarly thought. Now added to the standard IQ checklist we cite musical aptitude, both in terms of playing and learning, visuo-spatial ability, kinesthetic ability, interpersonal ability, social and emotional awareness, and many more forms.

Today, of course, there are a plethora of books that offer various takes on the multifaceted nature of intelligence. And in the sphere of intelligence science, where biological and genetic analysis has continued to hold sway, most scientists have softened their definition of what qualities determine intelligence. Even those most loyal to older notions have reported that making the most of your intelligence means marshaling its analytic, creative, and practical threads.

In short: to optimize your intelligence, you will need to boost much more than your IQ.

Still, the notion that we are born a certain way, and that our genes encode our potential, lingers with us as we move into the new millennium. The last twenty years have witnessed a gold rush in intelligence genetics thanks to the emergence of new brain-imaging technologies and genome science. Using a wide range of magnetic resonance imaging, or MRI, approaches, scientists have begun to observe how the brain works while in use. After studying the human brain while engaged in just about every imaginable activity, scientists now surmise that intelligence is a product of proper communication between networks and regions of the brain.[39] Geneticists studying the DNA sequence mapped by global initiatives like the Human Genome Project have confirmed this neural basis, identifying genetic markers associated with specific brain functions and behaviors.[40] This same research also suggests that when it comes to intelligence, brain size matters. Too big or too small, the brain is deficient if not defunct. Shape matters, too. The lace of a lobe is indicative of strengths or weaknesses in all kinds of abilities, with intellectual aptitude being the most critical.

Where has all this scientific advancement left us, in terms of truly understanding intelligence? What's the zeitgeist we're *thinking, knowing, enhancing* within today? Returning to our common understanding, we now regard intelligence not as one specific skill or gift but as a *set of skills* that can and should be improved. And yet, we also continue to regard intelligence as inborn and rankable, as though only a select few of us have been graced with superior brains.

Here is what I know to be true: intelligence is accessible to us all. And just as there are many ways to be intelligent, and many neural pathways that can facilitate it, there are also many ways to optimize your mind.

CHAPTER 2

Understanding IQ

Microsoft has long hired based on IQ and
"intellectual bandwidth."
—Bill Gates

Schools do it. Governments do it. Employers do it, too. Just about every social institution from education to medicine to the military administers IQ tests to assess the potential of those deemed worthy of admission. And despite the popularity of multiple intelligence theory and neuro-optimization, IQ is still viewed by most working in official capacities as the most accurate marker of a person's intellectual ability. A low IQ score can confer Disability status, while a high IQ score can confer Gifted and Talented status. As we move through our many social worlds in our day-to-day life, our intelligence is most often defined in terms of our IQ.

The problem is that, like our common definition of intelligence, IQ is an invention with many unscientific and highly problematic roots. And its statistical precision is a façade.

But before we can dismantle the system, we need to understand how it came to be. So, let's pause to consider the history of the IQ assessment and why it continues its stranglehold on our global societies. My hope is that, with a greater awareness of the intricacies of IQ scoring, we will finally be able to begin to challenge a practice that takes us away from the real basis of intelligence: learning from our environment.

Lies, Damned Lies, and Statistics

As we trace the origins of IQ, we must back up once more to the Edwardian Era. As Francis Galton set to applying his dear cousin's evolutionary theory to optimize human intelligence, he needed a way to measure it. Standardized testing became the number one tool.

For context, let's consider the limitations of the times. Modern biology was just emerging as a discipline. Most of the action was happening on the floor of the Royal Society, where Darwin and Galton held court over the top-hatted and tail-coated literati.[1] Darwin published *On the Origin of the Species* and won the Society's highest honor, the Copley Medal.[2] Galton then pioneered survey research and statistics, introducing the concepts of correlation, standard deviation, and logistic regression, using intelligence as his central variable.[3] He combined this new measurement science with

Darwinian theory to write the bible on eugenics, *Hereditary Genius*.[4]

Galton spent countless hours lecturing to his Society comrades that intelligence, like all else in the "cosmic order," traced the curve of a bell, introducing the now-famous "bell curve" into the scientific lexicon.[5] At one end of the curve were wizards; at the other dimwits, with a mass of mediocre minions in the middle. He established a mental testing center on the other side of Hyde Park, in South Kensington, where he created the first standardized intelligence test.[6] There he began averaging the scores to plot test takers along the curve.[7]

In 1886, Galton won the Royal Society King's Medal[8] and took intelligence testing on the road, spawning a multinational test-making campaign committed to intelligence quotients, soon to be known as "IQ."[9]

Galton's eugenics and genetics of intelligence spread like wildfire across the world, capturing the minds of the world's most influential people. Theodore Roosevelt, Winston Churchill, Oliver Wendell Holmes, and in institutional terms, just about every major university in Europe, the United States, and the colonized world.[10] The doctrine and metric were so powerful that they became the central justification for colonial expansion in the early 1900s,[11] and they held sway over the establishment of the newly minted field of genetics, with the first card-carrying geneticists being the staunchest of eugenicists.

The hegemony of IQ testing, governance, geopolitics, and

genetics was sealed. And the presumption that our individual intelligence could only be known by rank-comparison, rating us against others, became Rule No. 1 in science, education, and public administration.

The Quintessential Quotient

Though IQ testing began in earnest in 1904, when two French psychologists, Alfred Binet and Théodore Simon, developed an intelligence test for the French government to identify intellectually disabled children, little has changed in all these years.[12] The Binet-Simon test, now known as the Stanford-Binet Test (the world's leading intelligence test), cemented Galton's approach of creating an individual score based on comparing the test results to a reference group of test takers, in this case people of the same age.[13] When the French government[14] adopted the test in public schools and institutions throughout the country, no one could have imagined that the test would travel the globe and implant itself in nearly every nation in the world.

But within just a few years, IQ testing had moved to America, where it seeped into the recesses of society. Psychologist Henry Goddard, who was also working with developmentally disabled youth, translated the test into English and distributed more than twenty thousand copies of it across the United States, extending it beyond the realm of education to society writ large.[15] A fervent eugenicist, Goddard introduced the terms *moron, imbecile,* and *idiot* into the clinical lexicon, and convinced politicians to forcibly sterilize those with low scores.[16] After successfully instating IQ testing in the public school system, he set to work on using IQ scores to ethnically cleanse America. The U.S. government aided Goddard in establishing an intelligence testing center at Ellis Island.[17] All told, more than thirty American states legislatively adopted his compulsory sterilization program, which in 1927 was ruled constitutional by the U.S. Supreme Court.[18]

IQ testing became a cornerstone of screening for educational placement, career placement and compensation, clinical research and treatment, and admission into social clubs. In the U.S. alone, testing was performed in the military to screen millions of draftees to assign them to ranks and positions.[19] It was performed in education to find a system of tracking students based on "gifted" or "talented" versus "moronic" or "imbecilic" status. Books like *A Study of American Intelligence* convinced the government that immigration and racial integration were causing America to bloat with feeblemindedness.[20] As a result, in 1924 the U.S. government issued an Immigration Act to effectively ban all immigration from Asia and Africa, and to set country-specific quotas for immigration from Europe.[21] (This ban remained in place until 1965, when President Lyndon B. Johnson signed the Civil Rights Act into effect and began to dismantle segregation.)[22]

IQ testing was also a cornerstone of the Nazi "racial hygiene" movement and contributed to the deaths of at least a quarter million of the tens of millions who perished in the Holocaust.[23] Following from the "Law for the Prevention of

Hereditarily Diseased Offspring,"[24] German dictator Adolf Hitler signed into effect Aktion T4, a euthanasia campaign to administer a "mercy death" to "imbeciles" and others deemed unfit for life.[25] Victims were sourced from asylums, schools, and hospitals, with doctors and midwives mandated to report anyone suspected of a "serious hereditary disease" such as "idiocy." By the start of World War II, all children under three who met these conditions were automatically executed without consent of family, even newborns.

Race and IQ

By the 1960s, despite the beginning of civil rights organizing and the growing awareness of the atrocities that had been committed by the Nazis, the Binet-Simon scale (which had been revised by Stanford professor Lewis Terman in 1916, taking on its current name "Stanford-Binet") was more popular than ever.[26] Other tests, such as Raymond Cattell's Culture-Fair Intelligence Test[27] and the Wechsler Scales,[28] soon joined the Stanford-Binet Test in the pantheon of IQ measurement that continues to hold sway today.

But extending tests to the general public elicited a troubling pattern. White test takers were consistently out-scoring Blacks.[29] The average White test taker's score was, well, *average*. The average Black test taker's score was far below, worthy of clinical if not carceral intervention according to the law in countries like the United States. This was a moment in history when intelligence experts could have questioned the authority of IQ, yet far from inspiring critique, the "Racial IQ Gap," as it was called, inspired many to rally around IQ testing.[30]

Several influential American scientists went so far as to suggest that people of African descent were interminably dragging down the human race with their genetic disadvantage, as exhibited by low IQ scores. Stanford physicist William Shockley argued that Blacks should not be educated but rather sterilized to prevent further transmission of their broken genes.[31] Psychologist Arthur Jensen of the University of California, Berkeley, supported this argument in his 1969 *Harvard Educational Review* piece "How Much Can We Boost IQ and Scholastic Achievement?" wherein he argued against education for Blacks.[32] Jensen sought to add a modern touch to Galtonian eugenics, maintaining its emphasis on selective breeding and life-course tracking without the so-called "euthanasia."

A debate erupted in which faculty and students at Harvard and UC Berkeley, led by Harvard biologists Stephen Jay Gould and Richard Lewontin, protested this new IQ-based eugenics, identifying it as the racist practice it was.[33] In 1973, they and more than one thousand academics came together to issue a "Resolution Against Racism" in the *New York Times,* where they called for scientists, universities, and professional organizations to "eliminate classroom racism," "condemn and refuse to disseminate racism research," "expose the unscientific character of racist ideas so as to deny them the appearance of legitimacy provided by academia," and "organize and support activities to eliminate racist practices and ideas wherever they occur."[34] Gould and Lewontin began formulating counterarguments that they would eventually share in a series of critically acclaimed popular science books, such as *The Mismeasure of Man*[35] and *Not in Our Genes*,[36] outing the white supremacist motivations fueling modern eugenicists' shoddy research.

Unfortunately, these protests did not mark a sea change in our assessment of intelligence, but rather, they seemed only to kick up more debate. Just as I was entering my first year of day care, in 1979, the Racial IQ Gap controversy found its way to the courts. Up until this point, schools primarily administered IQ tests when a student appeared to be falling behind the median. A low score was grounds for placement in special needs "Educable Mentally Retarded" (EMR) classes. Yet, officials in my home state of California began noticing a pattern: educational tracking by test scores was inordinately placing Black students in EMR classes. In 1979, the state of California ruled that IQ tests "had a discriminatory impact on Black children," and therefore they could no longer be the sole basis for placement.[37] In 1986, the state extended the ruling to all placement of Black students in special needs classes statewide.

IQ for Good

While the IQ testing industry took a hit with the California ruling, it did not suffer for long, thanks to the work of an

American doctor hoping to shed light on the harm inflicted by environmental hazards. Harvard pediatrician Herbert Needleman had an inkling that lead exposure, not just lead poisoning, was critical to human health and development.[38] In 1979, just as the state of California was prosecuting its case against racist profiling in schools, Needleman published a paper proving that lead exposure diminished IQ scores. Needleman's study measuring lead levels in the children's teeth and bones found that higher levels of exposure correlated with lower IQ scores. He further found that low levels of exposure in utero and in early childhood, as the teeth were developing, lowered IQ and contributed to language acquisition delays and shorter attention spans. Needleman spawned a movement in science, medicine, and environmental justice to use IQ tests to prove the toxicity of environmental factors like heavy metals, pesticides, asbestos, and flame retardants that continues to this day.

That same year, Otago University (New Zealand) political scientist James Flynn also used IQ tests to detect inequalities in social environments.[39] His research into intelligence testing had found a universal rise over time in IQ scores. Flynn used this trend, thereafter referred to as "The Flynn Effect," to contend that IQ scores were not just malleable, but also environmentally determined. In an article in *American Psychologist* titled "Searching for Justice: The Discovery of IQ Gains over Time," he argued that the Racial IQ Gap reflected the fact that the average standard of living for Blacks in the 1990s was roughly equivalent to the average standard of living for Whites in the 1940s; one in which the latest in quality nutrition, health care, education, and living environment (even school facilities, sewage, and road quality) were absent.[40] He also showed that the Racial IQ Gap paled in comparison to the Generation IQ Gap between Whites born into different eras' standards of living, as well as between Whites living in the same time period but in different national contexts with large gaps in standard of living. IQ tests could unmask these inequalities.

Once again, there was a moment in which the international science community, and experts on intelligence specifically, could have questioned the veracity and authority of IQ. Though Flynn's work buttressed the value of IQ tests, it also pointed to their harrowing statistical flaws. Yet, once again, experts rallied to the cause.

In the late 1990s, the international intelligence studies community and the American Psychological Association (APA)—one of the most highly regarded professional associations worldwide and an authority on the science of intelligence—offered their stamp of approval on IQ tests, just as the scientific community was delivered an incendiary tract on genetics and intelligence.[41]

The Bell Curve: Intelligence and Class Structure in American Life, published in 1994 by Harvard psychologist Richard Herrnstein and American Enterprise Institute economist Charles Murray, argued that high intelligence—a heritable trait that could be accurately measured by IQ tests—was the best predictor of an individual's lifelong success.[42] Conversely, they proposed, low intelligence led to promiscuity, criminality, and overbreeding. As if Galton were clamoring from his grave, Herrnstein and Murray warned that a cognitive elite was forming and leaving the masses of average and below-average people in their wake, and that we as a society would be wise to let that elite live out its full potential for the good of the human race. They suggested that early childhood education programs like Head Start in the United States and family aid programs in the United Kingdom were a waste of resources. Their research, they said, offered proof that IQ travels along distinct racial lines, and that the criminality and poverty attributed to so-called "welfare queens" and "gang-banging thugs" "run in the genes of a family about as certainly as bad teeth." Pointing to the rise in crime in American inner cities and the concurrent decline in American academic standing on the global stage, they argued that it was best to face the facts and leverage them for the betterment of the *human* race.

The response to the publication of *The Bell Curve* was swift. Within months of the book's release, more than 400,000 copies were bought by people living all around the world.[43] Leading scientists from a diversity of fields—including genetics, psychology, and education—spun off their own studies, reanalyzing Herrnstein and Murray's data and publishing their counter-results. "Mainstream Science on Intelligence," a letter issued by a who's who of intelligence researchers in the *Wall Street Journal*, hit the newsstands in December 1994 with the message that data on the Racial IQ

Gap remained inconclusive because intelligence studies performed within racial groups showed substantial within-group variation.[44] In *Intelligence: Knowns and Unknowns*, the APA's Task Force on Intelligence similarly said that many people who shared a racial label lived in vastly different environments with great variation in standard of living.[45]

Despite sharing misgivings about *The Bell Curve*'s specific racial claims, the international scientific community buoyed the concept of IQ and validated the usefulness of IQ tests. They declared that intelligence was indeed a heritable trait best measured by IQ testing, and that intelligence was indeed structured by a bell curve on which races were generally positioned differently. What's more, they suggested that IQ tests were *not* culturally or racially biased, but rather were sound no matter which test you used. So, they agreed with Herrnstein and Murray that different racial groups tended to have different means and different curves, and that the Racial IQ Gap was real in some way. As they summarized, there was "no bias in the content or administration of the tests themselves," and "no adequate explanation of the differential between the IQ means of Blacks and Whites." IQ, gaps and all, was here for good.

What IQ Really Tells Us

By now, you might be thinking: *Fine, IQ tests aren't an accurate measure of intelligence—but surely, they must measure something?* Unfortunately, the jury is still out among intelligence researchers as to what tests measure with certainty, and there is little unity within the broader science community. Test makers claim that IQ tests show how well a person can answer one specific type of question, the relationship between shapes, numbers, and words (what science writer Siddhartha Mukherjee, author of the bestselling book *The Emperor of All Maladies,* calls a person's aptitude for navigating a certain kind of "problem space").[46]

But while test makers claim that this problem space and these kinds of questions directly measure a person's ability to reason, research on testing shows us that what they really measure is how familiar a test taker is with the language and format of the test, as well as how comfortable they are with the cultural reference points at the base of the questions.[47] Questions invoking Shakespeare, or teacups and saucers, score us on cultural know-how, which is a product of cultural membership. Are you an insider or an outsider? How prepped are you to decipher these words, signs, and symbols?

Data scientist W. Joel Schneider says that a good intelligence test goes beyond measuring how many "stupid facts" a person has memorized to tell us a person's visuo-spatial and auditory processing speed and quality, their short-term memory, and their ability to use new information.[48] (Most intelligence researchers agree with this, but many believe that these aptitudes are better measured separately and using tests that aren't loaded with cultural reference points in the first place.) But Schneider explains how hard it is to pose any question containing words, signs, or symbols that doesn't somehow require a test taker to have a cultural awareness, context, and connotation for those words and symbols. Ultimately, we use IQ tests to predict future action, a test taker's ability to reason with information in general, not just that which was presented to them on the test.[49] But even the shapes and numbers we take for granted as categorized and related in one single, universal way have proven to be different to members of different cultures. IQ tests have not translated universally after all. I mean, what does the word *circle* mean? What do you call snow? Is 2 x 2 the only and best way to ascertain a person's quantitative reasoning skills?

What's worse, test analysts find that in addition to scoring us on cultural membership, IQ tests score us on privilege. Test takers from lower socioeconomic backgrounds, those who live in poverty, those who lack proper nutrition, those who live in toxic environments, and those who endure toxic stress perform worse on tests. Meanwhile, test takers who enjoy health and wealth, and those who are able to eat nutritious foods, breathe clean air, and feel safe and undistracted perform better.[50] Even giving children simple vitamin and mineral supplements has boosted IQ scores in studies.[51] And as intelligence scientist Richard Nisbett has shown, test takers adopted into a family of a higher socioeconomic status often gain up to twenty IQ points.[52] That's enough to bump a person from one IQ bracket to another!

Work by intelligence researcher Angela Duckworth also reveals that IQ tests measure how much a test taker wants to do well on a test, their motivation level.[53] When you incentivize test takers to perform well with money, for example, their IQ scores go up significantly. This has held true for other cognitive tests that put a fixed number on our intellectual aptitude.

Further research into social motivations has likewise shown that cultural expectations and parental expectations also affect a test taker's score.[54] People who have been taught or coached to value intelligence in terms of IQ scores do better on tests. IQ tests therefore score us on our belief in the merit of the test itself, which is tied up in our cultural membership but also the privilege and social status of our parents, educators, caregivers, and relations.

Research has also shown that affective states, like what scientists call "task anxiety," affect a test taker's score.[55] People who are told that their brains are plastic and that they have an infinite possibility of scoring well have been shown to significantly improve their scores. People who are presented with reminders of their imminent failure, such as when researchers prime female and/or racial minority test takers with negative stereotypes about them, have been shown to significantly diminish their scores. Just being asked "what is your race?" can steeply downgrade a score for minoritized test takers who have been told their whole lives that their race is an impediment to a high IQ. In the end, IQ tests score people on their self-perceptions, their belief in themselves, and their internalized sense of value as set by the prejudices and misconceptions held in their culture.

So, when you hear that IQ correlates to how far a person

gets in life, aka income level, education level, job status, you can see that it is because the tests are a filter for just that. IQ tests confer higher scores on people who are already tracked to perform well in these areas. Tests do not actually tell us how capable a person is of thinking, taking in and using new information, making decisions, or solving problems. They tell us nothing about how variable a person's score could be if they were coached and consulted the way that some families do. Or how much they would improve if they had all the right resources, cultural and substantive, at their disposal their whole lives.

IQ Today

Many people, like me, took their first intelligence test amid these controversies. Yet very few would have known about the raging debates or known to question the tests themselves. Even Boomer educators, and the Gen X educators that followed, who fought against standardized testing and for student-centered learning in the late 1990s and early 2000s, would not have had the authority nor the tools to depose IQ testing. So, it has remained with us, as what ethicist Dorothy Roberts would call a "fatal invention."[56]

And, sadly, the court decision in my very own state of California did little to challenge the hegemony of IQ. California had the only statewide ban of testing, and only for Black students in California public grade schools from 1986 to 1992, when the presiding judge reversed his decision. The rest of the state, country, and world has permitted IQ testing in all institutions and industries. Microsoft and the rest are free to assess their incoming talent with this flawed filter.

In fact, despite the voices of so many crying "racism" and "bias," today IQ tests hold power over virtually every sector of society. In the United States where I live, the education system uses them to determine whether a person has a developmental disability and whether they are eligible for government aid. The justice system administers them to determine criminal sentencing. Private industry uses them to determine whether to hire and how much to compensate.

Now tests have found their way back into genetic science, as intelligence scientists scour our genomes for markers that can confer higher IQ. As we will soon discuss, this circularity has rendered IQ bulletproof. Now more than ever, the idea that intelligence is innate, fixed, and represented best by IQ is being built into our most cutting-edge technology.

CHAPTER 3

The Nature of Intelligence

"Know your genome."

"See you. Like never before."

"Did you read more books than your genes expected to?"

"Have choice over chance, the power to have healthier children."

The above quotes are just a few of the promises that DNA intelligence companies use to sell their products. We're talking apps, DNA IQ tests, and, since 2018, genomic prediction for the IQ of your unborn children. You can download your sequence to your phone, or link it through your 23andMe account, power up, and get forecasting. How is this possible?

We owe this technology to a burgeoning pursuit of intelligence genes. As scientists have racked up success for pinpointing genetic culprits for some of humankind's most devastating illnesses, some intelligence experts have turned to genetic methodologies to source variants that may be predictive of IQ.

Though these researchers have found little evidence to date, they continue to make bold promises that DNA variants will be found—it's only a matter of time. And companies continue to make bold claims that their technologies can work absent of proof. With these technologies in circulation, IQ now appears more rooted in DNA than ever. Meanwhile, IQ's faulty promise continues to eclipse the true nature of intelligence: learning from the moment and place in which you stand.

Genomics 101

Let's back up for a second and take a closer look at your genome so we can better understand what it means to capitalize on it. Your genome comprises long sequences of genetic material that code proteins.[1] These are made up of point mutations called "single nucleotide polymorphisms," or SNPs.[2]

To learn which SNPs matter to what end, you start with a phenomenon, something that's interesting from a medical, biological, or physiological standpoint. Like the New York City subway says, if you see something, then say something about it. In genetics-speak, you identify a phenotype—an observable trait of an organism (*pheno* being Greek for "showing")—and then you characterize it. In this case we flag intelligence, which we have then defined as an ability to learn, reason, or marshal your mind, and which we have characterized with scores.

Once you've identified and characterized your phenotype, it's time to ascertain its genotype, the organism's genetic material associated with it. There are a variety of methods for ascertaining a genotype, and they all have their place. Some methods focus on individual families and their genetic legacies. Others compare unrelated people or families, focusing on similarities and differences in trait holders, or siblings, or parents and siblings. Still others survey entire genomes of individual people who, for their unique biology, or just means and curiosity, are of particular interest.

The earliest human geneticists traced intelligence in family lineages. But as more became known about the genome in the late twentieth century, geneticists began using a candidate-gene approach, where they would map a gene known to be associated with the phenotype in as great of detail as possible.[3] This was easy when dealing with a single-gene, or "monogenetic,"[4] potentially deadly neurological disease like Huntington's, but harder when trying to study a multi-gene, or "polygenic,"[5] moving target of a trait like intelligence. Geneticists were at a loss until the early 2000s, when they invented a new technique for mapping polygenic variation.

When it comes to what geneticists call "common

variation," polygenic traits that all humans share but to different degrees, the genome-wide association study, or GWAS, has become the method of choice.[6] With GWAS, you can compare the genomes of unrelated people who are known to share the phenotype, again in this case intelligence, and yet share it to a specific degree.

Scientists call this a "case-control" method, because it compares "cases," or people with a specific measure of intelligence, with "controls," others who don't share that specific measure.[7] And scientists call the method "genome-wide" because geneticists use software to scour as much of the genome as possible, not just some specific area that may be already known to be associated with intelligence. So even if another study identifies a neurological gene variant on a specific chromosome, the GWAS will ignore that data and search agnostically for any information written into our DNA code. It's a powerful and elegant tool.

The only thing you have to watch out for is some other kind of shared trait that could be mucking up the data. Mutations are passed down from generation to generation, coursing the superhighways of our ancestral lineages. To make sure that you're comparing people who are similar in every other way possible, you must perform GWAS on people of the same background, roughly akin to what we call "ethnicity."[8] Even if you're working with pre-collected DNA, like DNA from a national database, there are simple algorithms that can help identify people who share ancestry to ensure that you're comparing apples with apples.

Genome and IQ

Geneticists have been using genome-wide association on IQ in recent years, turning up genetic variants that anybody with know-how can test for themselves or their offspring. For their studies, they compare scores of scores of scores, upwards of 100,000 and even over 200,000 IQ test results in some cases (in science, that's a lot of data!).[9] Looking at people with similar scores, like those considered super high (above 170) or just plain low (85 or less), scientists have cataloged hundreds of genetic variants that they believe are associated with IQ and may one day predict a person's intelligence level.

It would seem they have struck gold, and many personal genomics companies hang their hats on this data, citing it as evidence of the effectiveness of their intelligence-predicting products. Unfortunately, their claims are not so ironclad; there are grave problems with using genome-wide association on IQ.

First, researchers are comparing the genomes of people who share an imprecise, if not invalid, phenotype. It turns out that the scores of scores of scores used in intelligence GWAS have come from people who have taken *different* IQ tests.[10] Study investigators pool these differently measured scores because test makers assure them that any test given to any individual would produce a similar score—the tests are interchangeable. The scientific community has echoed this claim in the past, as we've just seen, maintaining that the tests all accurately work to the same end. But there is no proof that different tests produce identical scores, so the phenotype "high IQ" or "low IQ" is disputable.[11]

Second, their phenotype is corrupted because they are using biased tests, and very few of these biases are in test makers' control, as we have just seen. In fact, test makers have only answered calls to correct cultural biases, by ridding their tests of overt cultural content like references to Shakespeare.[12] They have actually worked much harder to keep tests the same as always, as a way to guarantee test interchangeability and comparability over time.

But even if there were heaps of proof that the phenotype "IQ" was rock solid, there are problems that are particularly vexing in the context of intelligence. Few genome-wide association studies unearth silver bullets for the population under study, let alone silver bullets that apply to all humans everywhere in the world. In medicine, only a few causal SNPs, like the ones associated with macular degeneration,[13] glaucoma,[14] and juvenile diabetes,[15] have been found. In behavior science, there have been *zero* causal SNPs found. In fact, in one of the field's largest and most noteworthy GWAS successes in studying human behavior, a study on educational attainment that found nearly 4,000 associated SNPs in over 3 million participants, the study still only could predict about 12 percent of what was going on with students' ability to make it through school. Eighty-eight percent of what was going on was not assessable with genetic measures![16] It would take thousands more SNPs (at the least!) to begin to have any predictive value for a behavior as complex as intelligence.

Another general problem that is exacerbated in the case of intelligence study concerns ascertaining a valid genotype. You can only claim that you have successfully identified a genetic variant for the ancestries you have included in your research. You most certainly cannot bring a technology to market if you have not performed GWAS on a diverse array of ancestries. Yet, about 90 percent of behavioral GWAS studies have been conducted on European ancestry groups alone, with almost 75 percent being conducted in people from Iceland, the United Kingdom, and the United States![17] This is particularly problematic given that companies sell their technology to anybody who will pay for it.

Scientists have only just begun using GWAS on diverse ancestral groups, carefully comparing the data of people who share ancestry but inhabit widely different environments, and the results have been inconclusive. Meanwhile, nongenetic research conducted by Flynn of "The Flynn Effect" has found standard of living to so thoroughly confound the data on heritability that there is no reason to continue seeking answers in the genome. His study of Black Americans has found that those living in the latter quarter of the twentieth century have experienced massive increases in their scores that can only be explained by social gains.[18] From Jim Crow conditions to circumstances similar to the present day, their averages have inched closer to the so-called "mean" of 100. And, more crucially, their changes have outpaced the general world population's gains over time (and over that period). In other words, the exact same group has outscored itself *and* the White Americans group it is so often compared to because its environment has improved. Genotypes could never explain this phenomenon.

Despite all these troubles, GWAS data is the reference point that the companies read your DNA against to make predictions for your innate aptitude and your future. This data also comprises the set of target variants for the brain-sculpting wave of the future. The engineers behind the curtain invest in the IQ paradigm, because they believe in it. Some claim that as geneticists enlist even greater sample sizes, the science will one day be so powerful as to be 100 percent predictive of a human being's intellect and potential.

Seeing Double

The sordid and specious history of the genetics of intelligence is both fascinating and dizzying. It almost feels like we're caught in some kind of conceptual whirlpool. Or perhaps more accurate, a conceptual sinkhole. Knowing full well that our perception of the skill set we call "intelligence" has changed little since the time of Darwin, we continue to dig in our heels with this troubling protocol.

When app creators, test makers, and software engineers are asked why their efforts are so singularly focused on the connection between DNA and IQ, they consistently offer the same answer: Because intelligence is not only measurable, but also *heritable*. About 80 percent of our smarts are built into our DNA. If you don't believe us, check the "twins."

Twin studies—the stuff of scientific dreams—are powerfully persuasive, and they entrance all in their midst. In these studies, researchers compare identical twins with fraternal twins, who share 100 percent of their genomes and 50 percent of their genomes, respectively.[19] Identical twins are considered biologically the same, while fraternal twins are considered as similar as any other set of siblings.[20] If researchers find the presence of a trait at the same rate in both types of twins, they determine the trait to be environmental. If they find its prevalence to be higher in the identical twins, they attribute that difference to genetics. The differential becomes what is referred to as a "heritability score."

The first twin studies ever conducted did not compare identical twins with fraternal twins, or devise heritability scores. They were introduced by none other than Galton, who charted similarities in intellectual aptitudes in twins and called it a day.[21] It was the bespectacled eugenicist Cyril Burt who popularized the idea that IQ was highly heritable, in the years leading up to World War II.[22] Comparing identical twins and fraternal twins across several different studies, he garnered an intelligence heritability score of approximately 80 percent. His rise to stardom came swift, as he was promptly elected president of the British Psychological Society and knighted by King George VI, becoming the first-ever psychologist to hold the title for his research contributions.[23]

Burt's fame flickered on, but his reputation took a nosedive toward the end of his life, when other twin specialists attempted to replicate his work.[24] It was curious that no matter what study population was used, and no matter its size, Burt always came to the *same exact* heritability score. There were huge gaps in his descriptions of not only how he went about conducting the research but also whom he studied. And the databases of twin pairs in existence showed there to be far fewer twins than he claimed to have studied. The numbers simply didn't match up.

Upon his death in 1966, Burt burned all his papers, making it impossible for anyone to verify whether the twin studies he purported to conduct actually happened at all. What is certain is that he made up the existence of many of the twin pairs in his research. Still, his claims left a permanent mark on the educational system in the United Kingdom. Citing his studies, British educators instituted a national policy of using IQ scores to determine which children were worthy of higher education. Standardized tests administered at the end of primary school and again after secondary school were used—and continue to be used today—as qualifying criteria for college prep and university.

After the Burt scandal, researchers began to create twins' databases in earnest, with trackable pairs of identical and fraternal twins. But soon debates arose over the validity of doing studies on twins who were raised in the same environment. Couldn't that inordinate similarity between identical twins be due to environmental factors like how they were parented, where they went to school, and how they were treated by people in their social worlds? There was no way to prove that it wasn't.

In the late 1970s, a new method was introduced: using identical twins who were raised apart, or adopted into different families, to compare with fraternal twins also adopted into different families. The Minnesota Study of Twins Reared Apart, or MISTRA, majority-funded by the notorious eugenicist organization the Pioneer Fund, was established to generate a definitive IQ heritability estimate that would corroborate its program of "race betterment."[25] Within a few years, it had achieved its goal. Surprise, surprise: MISTRA research confirmed Burt's conclusions.[26]

Since then, other twins' databases have been formed, most notable of all being the Swedish Twins Registry, which has nearly 200,000 twin pairs (75,000 of which are identified as identical or fraternal and 55,000 of which have contributed DNA), and the Twins Early Development Study, which has about 15,000 twin pairs (all of which are identified as identical or fraternal and 5,000 of which have contributed DNA).[27] Though heritability scoring has moved to these databases, with today's 80 percent official score echoed by Swedish Twins, no database other than MISTRA allows for the bulletproof method of comparing twins raised apart.

But is MISTRA really the holy grail it has purported itself to be? MISTRA, which closed in 1999, holds no genetic data (only behavioral data) on its eighty-one identical twin pairs.[28] And *none* of its twins were truly raised apart, separated at birth and kept apart unbeknownst to each other in completely different environmental contexts, such as varying socioeconomic and cultural backgrounds in different countries and kinds of communities.[29] Because MISTRA took place long before the internet boom, in a time before digital health and adoption records, to take part in MISTRA twins had to know of each other. Indeed, almost *all* grew up knowing that they had a twin living elsewhere.

Even more strikingly, while some of MISTRA's twin pairs had been put up for adoption, many had been partially raised together in the same home by the same parents![30] Their paths had diverged when their parents divorced. Or they had been put up for adoption after a period of being raised together. Of the twins that had been raised apart most of their lives, many had been reunited in their youth, and some had even come to live together again. Many also experienced a late separation and continued to grow up with one or the other parent; thus they had the same core value systems and socioeconomic and cultural environments shared by twins raised together. In fact, only twelve of the MISTRA identical and fraternal twin pairs were raised in different countries (a number far too small to create a heritability score), and even *they* experienced late separations and/or were separated after divorce and raised by their biological parents, ensuring continuity in environmental conditions.[31]

The Pioneer Fund and twenty-first-century eugenicists like Walter Kistler (sole benefactor of the Pioneer Fund and the Kistler Prize) have sunk millions into MISTRA with the aim of producing evidence that identical twins are truly identical, intelligence is almost entirely genetic, and there's no use in tweaking environments to help boost intellectual aptitude.[32] But MISTRA has offered no such conclusions about twins or genetics.

From Identically Different to Uniquely Similar

The relationship between our genes and our environment is complicated. There is no one moment when we are just genomes, because we develop in an intrauterine environment long before we see our first light of day. That is why using identical twins to fabricate genetic data in absence of genetic analysis is moot.

Here is something that may shock you: identical twins *are* never and *were* never identical.[33] That's right, there is not a *single moment* when they are absolutely identical. How's that possible? From the instant they split,, they are forming in relation to a shared set of resources, that is, their environment. There is no 100 percent genetics moment from a 100 percent shared genome that can be separated from that shared environmental context.

I learned this in the best of situations, living and working at a world-class medical institution where every possible technology was applied to track my twins' growth and development. I had an ultrasound *every other week* until the third trimester, when we shifted into weekly, then daily ultrasounds. And I can't even count the other tests and screens in between. Why so much surveillance? Because identical twins share resources. They share space, they share placenta, they share your attention, your care. They. Share. Everything. And that everything is *limited*.

Doctors name identical twins Baby A and Baby B, precisely so that they can ensure that they are both getting the resources they need. Baby A is the baby who first appears closest to the cervix; Baby B is the one floating above. From the outset, doctors establish baseline measurements to know which is which as they change position and develop. They also create a score for their size relative to one another. Any score of less than 25 percent variation is considered safe, while any score of more than 25 percent is considered riskier.

We were lucky. We learned that our twins had their own amniotic sacs. They appeared to have their own placentas, too. They were not at risk for twin-twin transfusion syndrome, where one fetus gets significantly more blood and nourishment, or selective intrauterine growth restriction, where one fails to grow and thrive. On the nourishment end, they could potentially get an even balance.

They did not. Just like with singletons in the womb, twins move a lot. They flip and flop and turn and spin. But as with single babies, the bigger they get, the closer they get to filling up the womb and positioning for their exit. In our case, Baby A occupied the bottom half of the womb for most of the pregnancy and Baby B straddled the canopy above.

At ultrasound after ultrasound, we noticed a pattern. Baby A had less room to move, while Baby B had the whole upper tier to himself. Baby B was getting bigger, too. Was it the backstroke? Was it his forward flips? We watched nervously as the score inched up to 10 percent, 12 percent . . .

And then they were here. Luca (A) first at 10:31 a.m. and Mars (B) at 10:32 a.m. Luca was three-quarters of a pound lighter than Mars, and his head an inch smaller. We sang. We cried. We felt a peacefulness like no other. They lay healthy in my arms nuzzling each other just as they had always done.

The hours passed and it was time to move to Recovery. As they wheeled me out, I caught a glimpse of the placentas and could not believe what I saw. There was only one! They must have fused at some point. And Mars's side was *enormous*. It was almost double the size of Luca's!

Alas, we had nothing more to worry about on that account. From now on, we could love them for their differences. We had a new MO, to raise them to be *unique*. And that is precisely what we began to do. As we dressed them to make our grand journey home, Luca in canary, Mars in chartreuse, we delighted in their own je ne sais quoi. In their cadence and call, we could already *feel* their personalities emerging and diverging.

We were eager to encourage each of our sons to establish his own identity apart from the other. We'd put one in corduroy overalls and the other in denim shorts, one in an aloha shirt and the other in a Fair Isle frock. We had already chosen names that had no relation to one another, illogical, discordant, incomparable in every way. All we had to do was give them significant one-on-one time with each of us and make sure they had opportunities to think, play, and feel apart from one another. They'd be a twin study's dream!

Punch line: we failed. True, we dressed them differently, gave them cacophonic names, and made Quality Time equal Alone Time. But they ate, slept, bathed, diapered together. It was virtually impossible to give each baby his own schedule, and why would we, when state-of-the-art pediatrics mandated that we feed and sleep them at identical intervals? We continued to be surveilled on promoting an identical growth standard in a series of measure-together doctor visits that lasted for months. The result? We were regularly awakened by a symphony of cheery gibberish and happy drivel, a product of their synced circadian rhythms. Today, they still fall asleep and wake within seconds of each other.

So, you see, environmental inputs come in all kinds of guises that become biology but are not genetics alone. A researcher could survey our identicals about eating, sleeping, pooping, or any bodily process and learn that they ate identically, slept identically, pooped identically, and so on. They could compare that to fraternal twins who weren't put on a strict eat-sleep regime, who didn't go in for countless pediatric visits to ensure that their growth and development was on par with each other's. And they could assume that eating, sleeping, pooping, etc. was almost entirely heritable. But unless identical twins are truly raised apart, they are being shaped by their environment. Their genes are expressing and regulating their bodies as a response to what their caregivers do. And that makes them more uniquely similar than identically different.

Enter CRISPR

False claims about genetic predictors of IQ might not be so widely dangerous if there weren't already market pressures to use genetic technologies to alter our minds and permanently improve the human race with a new kind of eugenics. In recent years, an even more powerful science has emerged with the potential to alter our intelligence at the cellular level: human gene editing. With gene-editing technologies like CRISPR (clustered regularly interspaced short palindromic repeat), geneticists have begun devising ways to rewrite our genomes to boost brain function. Some hope to create superhuman abilities, all with the tweak of a SNP.

Unlike personal DNA tests, which lie to the people who opt to take them and thereby harm a self-selecting few, CRISPR technologies affect our entire species. Though scientists can tackle specific cell types in an individual's body, their *soma*, so that edits are not passed on, the world community is fast moving toward modifying embryos or sex cells, at the *germ line*, so that changes can take effect before a person is born and be passed on to all future generations.

It wasn't always like this. When gene editing was first developed, the prevailing view in the science community, one held by the CRISPR elite, was that it should only be used as a tool to fight disease in living patients.[34] In CRISPRese, only "somatic" and only "terminal." Not on the germ line (inherited DNA that you pass to your children), not for a disease that wasn't life-threatening, and definitely not for intelligence. In fact, when Jennifer Doudna and other scientific leaders and government bodies all around the world convened the world's first international summit on human gene editing—where I spoke on the pitfalls of racially tainted genetic science—all agreed that we needed policies to prevent "designer babies," and especially attempts to weaponize gene editing by making alterations to the brain.

But by the time my own germ line was in full gestation mode as I was pregnant with my sons, the dominant view was that any life-enriching intervention should be game, and companies were beginning to explore edits for genetic variants associated with Alzheimer's, Parkinson's, and diseases that conferred intellectual disability, variants that could be mobilized for intellectual enhancements.

In the wake of the summit, I joined up with the first international gene-editing genome project, GP-write. I also videoconferenced with, phoned, and formally interviewed with many science writers, journalists, and filmmakers about CRISPR and its possibilities. People didn't just want to know what I thought about the science but also, given the slope we seemed to be slipping down, how I would use it.

As my hands lit on the familiar contour of Baby A's tiny elbow (or was it Baby B's knee?) on my belly, I confessed to my interlocutors that neither my husband nor I had any terminal illnesses in our family history. Sure, we had both had our fair share of cancer and heart failure, but in all cases there were clear environmental causes. But thinking about friends and acquaintances who had suffered from hereditary diseases themselves or with their loved ones, specifically those who lost children to inborn disease, I admitted that if we had any inkling of a terminal genetic disease, I would gene-edit to prevent it.

How about being good at things? Like playing music well? Being intelligent? I answered intuitively. My husband is a percussionist, a career musician, and he grew up farming clams and oysters with his family. I'm the opposite, a city girl who spent my whole life writing and dancing. The thing that joins us is our passion for learning and doing. To me, programming our twins' aptitude and interests would hinder that creative process. For the time being, I trusted life as I knew it. The CRISPR unknown seemed far too risky.

Then it happened. Someone else used CRISPR on their twins, using their unborn babies' cells to create the first designer babies. While the twins were edited to be resistant to infectious disease, the targeted variants also were implicated in intelligence research.[35] This was a big wake-up call. Despite our outrage at one of our own going rogue, we in the CRISPR community realized that there are no clear lines to draw between editing for disease and enhancement. By the close of the second international summit on human gene editing, a new consensus had formed. Somatic yes, germ line most likely, and enhancements of course.

At this point, I was pregnant again. Now optimization looked entirely different. As I lay prostrate on the gurney for the umpteenth time, ultrasound rendering the smooth cranial contours of Baby No. 3, I was getting the memo: gene editing in all its possible forms was going to overtake us like a tsunami. Eugenics was no longer controversial. There I was, researching and reflecting on what could happen, and it was already happening. With the move to germ line enhancements, I, a new and expecting mother, and we, the human race, no longer had a choice.

These realizations motivated me to more deeply question intelligence genetics, especially our heritability estimates and GWAS promises. The CRISPR juggernaut is already in motion, and a new eugenics is here. It is up to us to push back on the litany of false claims about the innate nature of intelligence, and to instead look to how we nurture ourselves and provide opportunities for growth to all.

CHAPTER 4

Nurturing Intelligence

The measure of intelligence is the ability to change.

—American proverb

Jorge was an engineer, a designer of gas and water lines in the bustling city of Bogotá, Colombia. One day a coworker spotted him working behind the meat counter of a butcher shop far on the other side of town. The coworker was perplexed. It was common knowledge that Jorge was a twin, but he was nothing like his fraternal-twin brother, Carlos. Anyway, Carlos wasn't a butcher, either. What was going on?

It wasn't Jorge. It wasn't Carlos. It was William, a farmer from a remote village more than 150 miles away. Jorge and Carlos would come to learn that they and their true twin brothers, William and Willard, had been switched at birth. These identical twin pairs had been raised as fraternal twin pairs with siblings totally unrelated to them in two highly differing families and two highly differing environments.

Jorge and William were identical twins. Carlos and Willard were identical twins. Their shuffle of environments made them the first true candidates for a "twins raised apart" experiment. Researchers at California State University, Fullerton, were eager to work with them, and in 2014, both sets of twins agreed to become subjects of study.[1] Finally, the scientists thought, they would be able to offer indisputable proof of the genetic basis of intelligence.

The men were compared on a range of mental and behavioral phenotypes, which showed that the twin pairs' varying environments had considerable effects on their outcomes. The most striking of these observations was that the urban brothers were completing graduate degrees and working in high-paying professions while the rural set had left school after the fifth grade to work in the pastures of Colombia's rustic countryside.[2] But what most surprised researchers was that the brothers' differences were evident in their DNA. Different genes had been turned on and off in their bodies throughout their lives.[3]

Over and Above the Genome

One of the biggest breakthroughs in contemporary science of the past decade has been our newfound understanding that our genes are not a fixed destiny, but rather that they exist, evolve, and express themselves differently in response to the environment they inhabit.[4] Genes can be turned off and they can be turned on. Their moment-to-moment environment governs their function.

There are actually several layers of "environment" encapsulating your genome. First there is the immediate cellular environment surrounding your strands of DNA. All of the cells in our body are specialized, and the function of a cell influences the expression of the genes contained within it.[5] Is it a lung cell, a heart cell, an eye cell? Does it help us breathe, pump blood, see the world? The "job" of the cell is an environmental influence.

Then there are its organ and organ system environments. In line with the examples above, we can think of the immediate environment of the lungs, which itself is a part of a larger pulmonary environment. Or the immediate environment of the heart, which connects to your vast circulatory system. We can envision your eyes and their terminals to the neural networks of your brain. Your DNA consists of the same instruction manual in each and every cell, but these shifting environments determine how your body encodes it. These environments determine which genes are needed now, for what, why, and how.

Beyond those immediate environments, all your organs are working together in the wider environment of your body, which is also signaling your DNA and telling it when to activate. Your lungs, heart, and eyes can't help you without your nerves, which reach into your core central nervous system, traveling up to your brain, just as they reach out to your skin, cilia, and mucous membranes at the surface of your body.

Even further beyond your body's inner corporeal environment, there is its outer ecology. There is the momentary environment in which your body is living, existing, acting. That environment is also signaling your DNA to encode in ways that help you fit into your environment, and help you understand your environment, and help you maneuver and manipulate your environment so you can make the most out of it.

So even if you had an identical twin who shared the exact same DNA and the exact same instruction manual as you, like Jorge and William, and Carlos and Willard, or like my own twin sons, your physiology and your life outcomes would already begin to differ at conception. Your unique environment would make you uniquely You.

The Epigenome

The science of an environment's impact on DNA is called epigenetics, and the focus of epigenetics is understanding the epigenome.[6] The epigenome is the sequence of code that binds to your genes (*epi* is Greek for "on top of" or "above")[7] and tells them when to turn on or off. Certain epigenetic sequences cause genes to silence themselves (also called "methylation")[8] while other sequences cause DNA to wrap itself so tightly that it temporarily can't turn on (what's called "histone," or "chromatin modification").[9] This switching on and off is fundamental to neurogenesis, your body's initiation and development of new brain cells. And it is essential to the gamut of neural processes related to intellectual ability, such as the tissue development and neural signaling that led Jorge and William, and Carlos and Willard, to differentiate beyond their surface appearance.

A popular way of explaining the epigenome is to liken it to the musicians who play a symphony. Your DNA is the sheet music. But you need your epigenome to put those instructions to sound. The most common way the epigenome works is to silence your DNA by encoding your genes to turn off. This can be helpful to your symphony in terms of telling your body when to start or stop playing a measure. But it can also be harmful if influences in your environment signal your epigenome to go silent when your body needs it to play.

The conductor of this symphony is your outermost environment, including lifestyle factors like pollution, diet, and exercise. Together they direct how your players read their music and decide whether your genes turn on or off. This influence can be both harmful and productive, depending on the health of your environment. Epigenetics has revealed that what you breathe and consume, and even how you think, all drive the function of your epigenome.

One of the most amazing things about the epigenome is that it is inherited.[10] Though it is imprinted by an individual's lived experience, it is passed from generation to generation along with DNA. That means that the environment your grandmother lived in, and the choices she made, have an effect on your present-day gene expression.

In fact, several longitudinal studies have revealed just how entwined present environments are with those of the past. The most well-known epigenetic studies are notable for illuminating the brain's neuroplasticity—its ability to grow and change all throughout life—and its responsiveness to the immediate environment, as well as its long-lasting degradation from encountering stress and trauma in the past.[11]

One notable example is a study conducted with survivors of the Dutch Hunger Winter, which occurred in 1944 when Nazis blocked food supplies to the Netherlands, creating a temporary famine. Those who were in utero that winter were born with epigenetic signatures that led to higher rates of schizophrenia, obesity, and diabetes, as well as shorter life spans. Researchers found that those markers were passed down to their children as well, evidencing that the stress and nutritional deprivation of a single season was enough to influence the biology of generations to come.[12]

Similarly, Holocaust survivors have been shown to have epigenetic changes affecting their brain tissue and leading to increases in anxiety, stress hormone production, and symptoms of post-traumatic stress disorder (PTSD).[13] Prisoners of war, too, have developed debilitating epigenetic signatures from their imprisonment in camps. Again, these changes were passed on to their descendants, who never experienced the specific stress or trauma they faced. The legacy of trauma on cognitive development is long.

While none of us can change the past, there is a hopeful upside to the epigenome's sensitivity to environment: just as your genome is not your fate, neither is your epigenome. It changes throughout your lifetime as a result of the experiences *you* have. So, you can improve your own symphony and even future generations' symphonies by making your environment a healthier one.

Healthy Environment, Healthy Mind

Epigenetic research focused on the brain has offered the greatest proof that our epigenomes are the complete control suite for our DNA from day one, and that factors like quality nutrition, air, and rest are required to initiate even our very first thoughts. Brain studies have shown that it is our epigenomes that regulate neurogenesis, or brain cell production, when we are embryos, signaling our brains to begin the unique process of cell division that leads us to have tremendously larger brains than our primate cousins.[14]

It is our epigenomes that regulate the hormonal genes that contribute to neurological dysfunction and decline later in life, too. Studies of maternal environments during pregnancy have further revealed that everything from a mother's vitamin intake to her meditation practice affects her baby's neural development and its ability to learn.[15] Even the father's consumption of vitamins and other "methylators," or silencers, has been shown to affect a baby's mental capacity, including the ability to perform intellectual tasks once born.[16]

But brain epigenetics have shown something even more striking: stress itself has a direct effect on intellectual processing. Studies on pregnant women have shown that late-pregnancy depression can alter newborn epigenomes in unhealthy ways, contributing to lifelong challenges with responding to the moment-to-moment mental inputs we receive and perceive as we move through our day.[17] Even paternal stresses incurred during a pregnancy can imprint a developing baby's epigenome, leading it to experience its world as stressful and, as a result, perform worse on intellectual tasks.[18] And though the few epigenetic studies focused directly on intellectual performance have used biased IQ tests to measure performance outcomes, these studies have proven that environmental stressors negatively affect test performance by producing harmful epigenetic signaling to activate stress hormones that cloud our vision.[19]

We all know what stress feels like. When we're faced with a stressor, our bodies brace and then rush with emotion. If the stressor is low-grade, we might relax or at least push our awareness of the stressor to the back of our mind. When the stressor is high-grade, our physiological stress response kicks in and our body fires off waves of chemicals and hormones in a delicate feedback loop that starts and ends in the brain.

This stress loop ignites with neurotransmitters like dopamine and epinephrine that alert your amygdala, your brain's emotional processor, to boot up your body's "HPA axis," its dedicated mind-body stress system that involves the hypothalamus, pituitary gland, and adrenals. These processes sharpen your mind's ability to use your memory and to take in new information that can help you to react.

Once your amygdala receives the signal that you are

undergoing stress, it alerts two parts of your brain to activate. First, it messages the prefrontal cortex, also known as the "control center" of the brain, to help you stay calm.[20] Your prefrontal cortex mines your memory for knowledge and images that can dial down the need to react, so that you can perceive your environment with less fear.

Then the amygdala messages the hypothalamus to rouse the pituitary gland to release the stress hormone cortisol into the body, which acts as a stimulant.[21] Your adrenal glands flood the body with cortisol, igniting other organs to react to the stressor. Muscles spark, breathing quickens, heartbeat hastens. Your mind stops reasoning and taking in new information. Instead, it gathers images and ideas that enable you to fight or flee.

Somewhere around the thirty-minute mark, the adrenals must stop distributing cortisol and the cortisol in your body must begin flowing back to the brain. The brain must close the loop. The body must return to normal, so that you can regain access to the parts of your brain that you need to reason and learn. It takes sixty to ninety minutes from the onset of the stress response for your body to return to its baseline state.

Given the timing of it all, researchers have found that a mild, acute stressor that occurs just before a related event (think butterflies in your stomach before an exam) benefits from that initial wave of chemicals and can potentially enhance your cognitive abilities. But when cortisol levels remain elevated for too long, they start to damage the brain.[22]

What's worse, series of unclosed feedback loops lead to permanent loss of cognitive function and eventually may lead to degeneration and intellectual disability.

There are short-term and long-term repercussions of stress on the brain. In the short term, the elevated heart rate and breathing that increases blood flow and oxygen in the body also diverts energy from the parts of your brain that enable deep cognition, attention, and memory.[23] A mass of psychological performance studies has proved this, showing how people fail at everything from visuo-spatial tasks and memory recall to calculation and comprehension under even mild acute one-off experiences of duress.[24]

But stress is rarely a one-off event. Most of us experience chronic stress throughout our day, every day, every week, all year. Researchers have thus shown how in the long term, enduring stress creates structural changes in the brain, leading to brain atrophy and neurogenesis disorders.[25] When your HPA axis is constantly stimulated and the brain is constantly flooded with cortisol, the system itself goes haywire. The brain looks for ways to stop producing cortisol, diverting resources from other regions and structures of the brain. Measurements of neurons in brain regions responsible for higher thinking, learning, and memory—like the hippocampus—take the biggest hit, as studies showing reductions in their numbers and volumes have shown.

In the long term, stress can also have the devastating repercussion of triggering neurodegenerative disease. Spiked cortisol levels have been shown to lead to the early development of what's known as "mild cognitive impairment," a disease defined by memory lapses, having trouble coming up with words and ideas, losing things, and forgetting to show up for planned events.[26] Prolonged stress may also lay the foundation for dementia, and 80 percent of those with dementia go on to develop Alzheimer's disease.[27]

We know that acute stressful life events, like going through a divorce or losing a loved one, often trigger depression. But contemporary neuroscience has shown that acute stressful events can have lifelong and even intergenerational effects on the brain. A large body of studies has demonstrated that people who experience highly stressful life events are more likely to suffer from cognitive dysfunction and decline. Studies of pregnant women have also shown these events to negatively impact gestation, leading to premature and low-birth-weight babies.

Up until now, I have mostly focused on the effects of acute stressors. But today's research shows that experiencing chronic stress is far more debilitating. The chronic stress of high-pressured school or work settings, painful or abusive relationships, or discriminatory social environments weakens the brain's ability to close the stress-response feedback loop, thus inducing an endless cycle of intensifying stress responses.[28] Chronic stress also promotes inflammation across bodily systems, adversely affecting the brain and its functions via circulatory and cardiovascular degeneration.[29] That is why we see stress being linked to just about every disease of the brain, from mental illness and neurodegenerative diseases to cancers and immune disorders.

Today there is a rapidly developing field of study within epigenetics devoted to studying the wear and tear incurred from stressful environments. "Social epigenetics" examines how factors like education, nutrition, housing, exposure to pollution and toxins, child-rearing practices, parental health, and education all program and reprogram our minds throughout our lives.[30] Studies of crumbling school systems, deprived neighborhoods like the under-resourced communities studied by James Flynn, and inadequate healthcare facilities show how harmful environments get "under our skin" and into our brains as we face an onslaught of stressful living. This work has motivated many scientists to call for increased attention to the policies that determine the quality of our social environments. Those focused on intelligence have been especially vocal about changing education systems, noting that even minute strides toward equal access to quality environments can inspire lasting improvements in people's intellectual ability for generations to come.

Nurture Your Intelligence

The neurological and epigenetic impacts of stress show us that being born with a certain sequence of genetic code doesn't guarantee anything if your environment isn't set up in a way that allows you to take advantage of it. Likewise, editing a certain sequence of genetic code won't do anything to change your circumstances, which are the key determinant of your mental clarity and intellectual health. While epigenetic CRISPR may one day cure those with congenital neurogenesis disorders like Tay-Sachs, macrocephaly, or fragile X, it won't change anything for those of us in the general population if our stress triggers remain the same.

This window into DNA function helps us see clearly that the epigenome is the "on" button for your thinking mind. Lucky for you, your brain is genetically hardwired to think intelligently, to allow you to use your mental faculties to get the most out of your environment. Your sheet music holds instructions to do just this, to play your heart out and make the most of your unique sound. But you must ensure that nothing is standing in its way. You must hire the right conductor to lead the charge. You must fine-tune your players to live up to all you can be.

How then do we improve our epigenetic outcomes? We must apply tactics to reduce stress and stimulate our brains in healthy ways. While there are many ways to battle stress and promote mental health, I want to focus on cognitive strategies that enable our brains to do the work we need them to do, in science-backed ways that don't require a high cost or invasive interventions.

In Part II, I will introduce a framework that helps us optimize intelligence by using our minds to connect with, learn from, and improve our environment. I will offer ideas on how to initiate an immersive learning-as-you-go outlook that can help anyone seize the myriad growth opportunities their environment provides.

There's a lot we don't know about how to acquire intelligence, but one thing we know for sure is that a healthy environment shapes a healthy mind. We must work with our nature to nurture it.

PART II

Nurturing Intelligence

CHAPTER 5

The Growth Mindset

There is no failure, only feedback.

—Robert Allan

Understanding that intelligence cannot be reduced to a score that ranks us against other people, or a handful of genetic variants, helps us to begin seeing how we can truly optimize it. Intelligence is a dynamic process by which we interact with our environment to extract knowledge from it to improve our lives. Everybody, no matter how smart, academic, or "neurotypical" they seem, has this capacity and can tap into it.

Understanding that intelligence is also more than mere intellect, and is mediated by stress and our mental and emotional health, additionally helps us to begin building a better environment for our minds. Because our brains and bodies are constantly responding to the social environments in which we work, learn, and live, we can question how enriching those environments are. We all have a responsibility to improve the impact of our external life on our internal life, of our environment on our epigenetics, so that we all can be empowered to hone our potential.

Key to flipping the script is optimizing your own knowledge of a human being's inherent neuroplasticity—our brain's ability to grow and change through life. A person's intelligence isn't something fixed at birth because the brain is constantly growing and changing. A person's intelligence pivots on how they use their mind to forge connections with their environment, and especially how they use their environment to forge connections within their mind. By thinking and talking about intelligence as fluid, and the processual nature of learning from our environment, we can make our homes, workplaces, and spaces of learning nurturing ground for our intelligence.

Upgrading the Brain

More than a century of human genetic research has proven that the human brain is plastic: it is adapted to develop and grow through the quintessential human activity we call learning. The brain is fine-tuned to help us take in new information, compare it with the information we already have, and take action to improve our lives. This cycle of input and output is what the brain was designed for, and it is this very act that allows our brains to generate new neurons and make new connections. Our responsiveness to our ever-changing environment feeds the plasticity of our brains, literally rewiring and enhancing them, giving them an upgrade as we move through our lives.

It used to be thought that brain cells only formed when we were embryos, and except for the brain's eventual atrophy in old age, scientists saw the brain's pathways and networks as fixed throughout life. Yet, we now know that the brain is constantly growing, developing, and improving itself beyond birth, childhood, adolescence, and early adulthood.[1] Brain signals in the form of electrochemical flows trigger our stem cells to commence neurogenesis and make us new brain cells. This then spurs the brain to rewire its connections and form connections better attuned to our life's needs.

Learning is the spark that lights your brain up and gets its signals to flow. Repeating a mental task and asking your brain to retrieve memories around an activity literally ignites your brain to strengthen its neural pathways.[2] And when you challenge yourself through thinking and practicing in a certain way, say by doing math equations or by painting or meditating, your brain begins to build beyond its former pathways to help you perform that particular mental activity better. When you stop doing that form of learning, your brain eventually replaces those pathways and networks with other ones. It "prunes" the cellular connections down so that you can invest your neural energy where you now need it most.

Learning *new* things creates an even bigger spark. Forcing your brain to respond to your environment in new ways tells it that it is time to create new pathways and new cells. So, in addition to strengthening old ones, reworking them, and building beyond them, your brain grows new material to make new networks.

Life Is Learning, Learning Is Growth

On a macro level, learning leads us to creatively find ways to work together to live in harmony with others who share our environment and to work together to live in harmony with our environments themselves. Just think of the power of learning to read and what collaborative possibilities reading and writing open up for humankind. On a micro level, our brains are literally engineering and reengineering us to make our environments better work for us, making cells collaborate and build up networks as if it were a microcosm of the macro level. Every DNA strand, every cell in our body, every activity we engage in is an expression of this most basic fact of life.

It is popularly said that learning is a kind of power or an investment, as if some have more or make more from it, but truly, learning is the basis of life. Your brain gives you all you need to gain knowledge about and mastery over your environment, and your environment gives you all you need to nurture your brain. It is how our brains function and what we all do. *Every moment is a learning moment.* Every moment provides you with an opportunity to know, maneuver, and manipulate your environment. There is always a message, no matter how routine or mundane. It is up to you to seize it. And though some moments are breezy and others are tempestuous, you are always adapting and evolving. You are always potentially one step closer to fostering creativity, compassion, and contentment.

Sure, you will prioritize learning about some topics or subjects over others in particular moments and parts of your life. I spent the first third of my life focused on acquiring street wisdom in my neighborhood, studying math and history at school, and training for a professional dance career, only to spend the second third of my life focused on learning how to have healthy relationships, build my community, and become an expert in social science. But these topics and subjects didn't define my intelligence. Rather, it was my day-to-day learning that enabled me to use my immediate environment to think better and live smarter and be more socially aware, regardless of the overarching themes I believed were most valuable.

Likewise, your leaders, managers, teachers, and colleagues will often require you to focus on aspects of your momentary environment to fit their (or an institution's) agenda. In my present stage of life as a professor and public intellectual, I am constantly being asked to learn, teach, and write about new areas of expertise that can better serve a fast-changing student body and public. I am also required to do specific tasks for my university, my profession, and the broader scientific community every single day. But these areas of expertise and job tasks do not define my intelligence, either. Instead, it is my continual conscious engagement with my environment that drives my brain to grow and fortify.

So don't be fooled into thinking your smarts are made by checking academic boxes or working toward performance endpoints and goals set by others at work, such as when you reach a professional milestone or solve a problem that is plaguing your firm. Sure, formal learning and problem-solving can help us become more knowledgeable and disciplined, and performing our mastery of various areas of expertise can help us do our jobs. But in the end, it is only by capitalizing on the learning moment and working toward your own personal goals, no matter what externally imposed task is at hand, that you can you achieve your true potential. That is how you live intelligently.

This is why it is so important to reconsider how we define "intelligence," collectively and individually. We must reject the notion that intelligence is a zero-sum game. We must replace that notion with a processual, growth-oriented definition that takes into account neuroplasticity.

With this in mind, ask yourself: Do I consider my intellect to be fixed or fluid? Is my motivation for learning externally driven—am I trying to achieve a goal set by someone else—or is my desire for knowledge something that feels essential to my being? What values and assumptions about intelligence have I inherited from family and from culture and society? How do these ideas inform, and possibly limit, my confidence and ability to grow? Am I identifying the continual growth opportunities provided to me by my environment? These questions can help you retrain your way of thinking about intelligence.

Setting Yourself Up for Success

To optimize your intelligence, you will need to develop a growth-oriented mindset that extends far beyond your intellect, beyond your milestones, your achievements, and your expertise. You will need a mindset that redefines success in terms of your commitment to encouraging your brain's innate neuroplasticity. And you will need that mindset to be reflected in your wider world—in the education and work systems that you move through, and in the social and political structures that provide the ultimate nurturing ground to you and others.

But before you can do that, you will need to assess your current attitudes and beliefs about intelligence and learning, and replace any notions of achieving a measurable uptick in intelligence with a new aim of embracing growth and development. So, let's take a moment to open up and unpack some of our biases about achievement that all too often hold us back from actually attaining it.

The desire to live a life marked by achievement is understandable. On the surface, it would appear that charting a course for performance outcomes like getting exceptional grades, going to an Ivy League school, or climbing the ladder in your given industry could only improve your life. The me of my childhood dreamed of being a ballerina on the international stage, and then later a Supreme Court justice determining the fate of my country. The me of my early adulthood set her sights on being a venerated academic with a big public impact. When I began my career teaching in one of America's most popular Ivy League schools, Brown University, I thought I had arrived at success.

However, as I learned, there is a cost to such high achievement, and it is one that, as we've discussed, has significant downstream effects on the mind and body: stress. For example, while high grades and test scores would seem a very worthwhile achievement to work toward, global studies have shown that academic performance–related stress is linked to poorer quality of life and well-being.[3] Students under high academic pressure are more likely to suffer from anxiety, depression, and insomnia. The same holds true for adults who are in high-stress jobs. Global studies have shown alarmingly high percentages of job performance-related stress in adults, with some countries reporting the majority of workers experiencing daily distress.[4]

This phenomenon is most clear in high-achieving environments such as the Ivy League towns and research and development hubs of the world in which I have lived and worked (Princeton, Palo Alto, Manhattan, San Francisco . . .), where youth report off-the-charts anxiety and workers report feeling stressed all the time. One study of a competitive Silicon Valley high school revealed that 80 percent of students felt moderate to severe anxiety, with over 50 percent reporting severe depression.[5] Another study of the Valley's workforce reported that over 70 percent of workers were experiencing burnout and work-from-home exhaustion during the first years of the pandemic.[6] A mass of workplace research in these environments has consistently found that high-pressure work environments create negative and abusive cycles of interaction that can lead workers to experience PTSD.

Even more disturbing are the teen suicide clusters that have plagued communities known for high scores on school standardized tests.[7] These are often environments of relative affluence where youth are acculturated into a score-based definition of success, and where the end goal is to be in the top percentile of the country so that you can get into an elite private university or college. Longitudinal "life course" studies that have followed students from high-achieving schools into old age have found that, regardless of their level of affluence, these students are more likely to experience mental illness and to use substances while in school as well as after graduation. One study found that their alcohol and drug use was two to three times the national average in college and in the years after graduating from college when their work and careers commenced.[8] Yet, some never make it that far. By middle school, kids in areas like these may feel so much pressure to live up to their community's goals that they just want to give up. Children as young as nine years old who have the highest marks in school have succumbed to the pressure to achieve. Indeed, suicide is the second leading cause of death for youth in "WEIRD" (Western, Educated, Industrialized, Rich, and Democratic) countries like the U.S. Worldwide, it is only topped by HIV and road injuries as the leading cause of death for young people.

Even in suboptimal environments where schools are

underserved, students are plagued by standardized testing and score-based academic stress. In fact, failing schools are more likely to be on an all-year testing treadmill than their well-funded counterparts in order to prove their worth to the education system. So, in addition to the stresses of poverty that prohibit learning in these neglected environments (something I discuss further in Chapter 8), low-performing environments equally pressure students in unhealthy ways due to our score-based ranking systems.

These numbers show that the stress to achieve and its consequences are not reducible to personal problems. The pressure to score and rank high is endemic on a cultural and community level, and increasingly on a global one. While you can't control all aspects of your environment, you can question and rectify the things you hold true about achievement, so that you are setting yourself up for true success according to the rules of the most fundamental aspects of life.

Growth Not Scores

We live in an era where data reigns supreme. Bits and stats are constantly pouring in, and we are measuring ourselves and being measured by them all day. The average American grade school student takes more than one hundred standardized tests over the course of their K–12 education.[9] Meanwhile, workers in industries all over the world are managed with performance software that tracks their output in real time. Our sayings and doings are quantified at a rapid clip.

The problem with scores is that they are mere snapshots. When administrators hand out IQ tests or other forms of standardized tests, they analyze the results of those tests as stable fact. A person is thereby branded as possessing a certain capacity, as if they are innately only so "good." And very often, that person comes to see themself in those terms. To this day, I can recall the pride I felt as a small child scoring in the top percentile of reading and writing on our standardized tests, and conversely the shame I felt when I scored far lower in vocabulary on those same tests. I believed I just wasn't good at learning and understanding words, when all I really suffered from was a lack of exposure as a result of growing up in a home where English was a second language. My own belief about myself as indelibly best at certain areas of learning and indelibly worst at others became a self-fulfilling prophecy that affected my interest in learning the very subjects for which I needed help.

What's worse, scores are built-in ranking agents that deem some to be winners and others to be losers. They purport to tell us who flourished and who failed. In reality, they produce a zero-sum game that advantages those who have had access to test information, practice, and coaching and consulting, not to mention safe and healthy living environments, good nutrition, and the like.

A growth mindset rejects this bean counting and the corresponding "audit culture" that encourages us to view and value ourselves in terms of indicators and benchmarks. It instead sees humans as neuroplastic, as malleable organisms with brains that are constantly growing and changing.[10] It sees life as a learning process. It envisions us all as inherently intelligent in that we all can seize the opportunities that are constantly presented to us. A growth mindset sees us on a ride of life.

In this journey, nothing is permanent. There is only development. Humans by definition, that is, by the evolutionary adaptations that led us to this unique form, are lifelong learners. Thinking in terms of our neuroplasticity, and our journey toward mastery of whatever we put our mind to, can help us hitch up and ride in style, with the wind at our backs, propelling us to greater understanding, wisdom, and knowledge.

Seeing and Teaching Growth

I have adapted the term "growth mindset" from the work of psychologist Carol Dweck, which focuses more squarely on how people's assumptions about their own intelligence affect their performance. Dweck initially introduced the term in the early 2000s to reveal how some concepts of intelligence can hinder a person's ability to learn. Dweck's mandate was to instruct people on the neuroplasticity of their brains and then to teach people to see that learning is a long road with pitfalls that they *can* and *will* overcome.[11] Since then, there have been many helpful adaptations of Dweck's basic protocol.

For parents and caregivers of children, you can begin with talking about the brain and how learning literally builds the physical connections within it.[12] Then you can move on to using relatable examples (like learning to read or learning to dress yourself) to teach how learning is a process.[13] It is imperative to share your own process of learning by reflecting on the feelings you have felt when hitting a roadblock or making a mistake ("First I was excited but then I was frustrated when . . ."). I have read excellent junior picture books on neuroplasticity to my kids starting at age three. There are also a number of growth mindset books for children that focus on the malleability of brains and overcoming learning obstacles that can help prompt you to begin the conversation.[14]

But most helpful is to share all these experiences with your own journey as they come up in your everyday life. For example, when you make a mistake or when you struggle to achieve something, you can speak aloud about how it is making you feel and how you are facing your obstacles and learning from them. When I am cooking with one of my kids and I add sugar instead of salt to a recipe, I tell them, "Ooph! This is not going to taste right. I'm bummed that I wasn't looking at the page and have to start over. But I'm glad we're gonna get this right! Next time, how about I read the recipe out loud to us as we go?" Another example is when a child is engaged in a challenging activity. Remind them about your struggles to become proficient. My kids are learning to swim, so I like to remind them about the many different swim teachers I had at different times in my life—first my dad, then my friend's mom, and then a series of camp counselors, who all helped me hone my ability beginning at age four and continuing through my teens. Becoming a proficient swimmer wasn't something that happened overnight. But I kept at it, even when I felt inadequate. I also like to tell them about my mom, who almost drowned in the Indian Ocean as a child and didn't try swimming again until she was in her sixties. Now when we go to visit our family in Indonesia, she organizes swim instruction so that she can join us. I tell them what she has told me: that she will always feel scared when she first dips her toes into the water. But knowing that she is proceeding safely helps her overcome that fear so that she can continue on her journey of learning.

For teachers, educators, and others who find themselves in a trainer role, you can also begin with a lesson on neuroplasticity (albeit more detailed) and then structure your learning and training activities so that they provide room for reflection on the processual nature of learning.[15] For example, in place of the "teaching to the test" (in other words, instruct so that you can evaluate them) protocol, you can give a lesson and then allow the learner to have time for open-ended inquiry and problem-solving. It is important to give the learner challenges and obstacles to overcome, and to talk through the process of overcoming and the feelings that it brings up. In my classes, I have replaced quiz and test time with what we in higher education call "problem-based learning." I lecture for ten minutes and then give students a real-world problem to solve together that follows from the lessons I taught (I'll share more on this form of collaborative learning in Chapter 7). I allow them to do the actual work of sociology, which includes adapting sociological concepts to the real world, using them as a lens to draw out and explain systems like gene-environment interactions, biotech industries, pharmaceutical pipelines, and government regulatory structures. I also shadow them and commend them for their use of particular strategies, their engagement with knotty ideas, and their effort to learn—this as opposed to praising some for being smart or scoring high on a quiz. I let them see for themselves how practicing sociology is just that: a practice. I prove to them that to achieve in my classes, you have to exercise your sociological imagination and perform different approaches. You have to try, fail, get up again, and practice, practice, practice.

Just as with kids, with learners it is critical that you model the journey to mastery. For me, teaching and mentoring are indistinguishable, and I treat all my students as apprentices. I talk about my own education, my own career, and my many mistakes along the way in just about every lecture that I give, and I do so even though 99 percent of my students will not follow my footsteps into science and academia. I model the processual nature of my own path to mastery of my subject because my job isn't to test, rank, and brand the "gifted" as academic elites. My job is to inspire my students to think in new ways, to capitalize on their brain's neuroplasticity, and to grow from what they are learning.

In the work environment, teaching on neuroplasticity and teaching and modeling growth is just as important.[16] And it starts with leadership. If you are an employer, leader, or manager, try talking openly about your own struggles and journey within your organization, but also in your broader career. Make sure that your workers know that you see their current position as temporary, that they have room to grow, and that you believe in them and their growth. Explain to them how working in a particular role is an opportunity for gaining mastery and expertise. Identify what they will be learning and detail how that can propel them to new roles and positions within your organization and beyond. Treat their progress as personal, not just organization-serving. In my work, I begin each quarter by sitting down with my team members and discussing long-term career goals. From there, we work backward to ensure that the tasks and responsibilities they have now will lead them to those long-term goals. We talk about new and ongoing assignments in terms of the learning struggles that they will face in the coming months. I help them strategize how to possibly overcome some of those struggles, using examples from my own history.

In the work environment, it is also critical that you personalize your mistakes and discuss them in ways that elaborate on the emotional nature of learning and the satisfaction of turning a failure into an opportunity.[17] Though I work alone from my home office much of the time, I recount my failures, my feelings, and my process for overcoming hardships when I meet virtually with my team. When I am in person with colleagues, I pay particular attention to those moments when I feel embarrassed. I use them as little red flags, or signals, to reflect on what I feel I am doing wrong and to share aloud my own mistakes and my process of surmounting difficulty. I let my colleagues and my subordinates know that I do not aim for (nor do I value) perfection. Rather, I aim to do my honest best, *especially* by seeing and teaching when I can course-correct. I value the iterative process, the day-to-day practice of my profession, and the concerted effort that allows me to move forward on my personal trajectory.

Verbalizing and modeling a growth mindset must be met with a workplace culture of growth. For example, when an employee makes a mistake or fails to meet their performance endpoints, how does management respond? Do they publicly take the worker to task? Do they fire them? Or do they talk about the error or failure as a learning opportunity? In my work, my colleagues and team members are often overwhelmed with work and unable to meet our deadlines. And even with my Fellow Parent or Leader or Mentor hats on, I am frequently met with disappointment and frustration at others. But instead of falling into a blame-and-shame routine, I reflect and reframe. I ask myself how they might be feeling and ask myself how this incident fits into *their* trajectory, *their* process, and *their* struggle. Finally, I ask myself if this is a big enough deal that it merits discussion. If so, I find a way to talk it over with them so we can reflect on how to overcome the mistake and how to get past it in a way that bolsters all our learning and mastery. These steps are essential to building a better understanding of one's own intelligence. They can also set us on a better path forward to meet our personal goals for achievement.

Fluid Not Fixed

Seeing intelligence in terms of fluidity—and in terms of brain growth, personal growth, and mastery of knowledge—holds numerous benefits, while seeing it in terms of fixedness—as with the permanence of a score—holds numerous detriments. Those who believe that their intelligence is something they are born with, something fixed at birth, tend to fear and avoid the unknown. They see success in binary terms and view their own abilities as interminably limited. They view the horizon at their own edge of knowledge with a sense of doom. As a result, they experience feedback on what they don't yet know as criticism and with a deep sense of failure. They believe that effort is pointless, so they give up easily. At the same time, they are plagued by a desire to prove that they are naturally gifted. Their motivation is externally driven with an eye toward external validation, such as winning awards or achieving a certain level of recognition. They see themselves only in relation to others, to their comparative ranking, and perceive those who score higher as inherently more valuable. They are threatened by the perceived success of others. They see little value in learning.

By contrast, those who believe that their intelligence is incremental, something that develops throughout their life, see life as full of learning opportunities. Mistakes and setbacks are challenges to improve knowledge. Feedback, even criticism of blunders or weaknesses, is encouragement.[18] So too is others' success, because it shows a better way forward and highlights opportunities for course correction. Those possessing this kind of mindset see learning as the most essential part of being human. They embrace it as an exercise regime to support their brain.

The difference in outcomes from these mentalities is striking. Nearly a quarter century of research in fields as diverse as psychology, neuroscience, and intelligence has proven that adopting a growth mindset leads to better knowledge acquisition, learning experiences, and test and performance outcomes.[19]

For families, a growth mindset has shown to improve outcomes for parents as well as children, especially in the areas of parental and child well-being and kids' test scores. Comparing parental and child mindsets has shown that children adopt parental mindsets, and those who see their own lives in terms of scores, endpoints, and failure transfer that to their kids, thereby slanting their outcomes downward.[20] Even very young children have been shown to hold fixed versus fluid notions of their own intelligence, and as a result have performed better or worse on exams and mental tasks. The research in parenting suggests that children who are exposed to a growth mindset view their world with hope and appreciation, and that perspective allows them to embrace learning from the outset, which further supports their own neuroplasticity.[21] With a love of learning for the sake of learning, and with the experience of the joy of the process of learning, young children are able to develop healthy patterns of knowledge acquisition and avoid stress.

The education systems I have worked in, where the score mindset dominates, also illustrate the point nicely. An abundance of studies has revealed that scoring people with tests is not only a terrible way to foster learning, but also a terrible measure of how much people have actually learned.[22] What's worse, scoring squanders the myriad learning moments in an academic year and replaces them with stressful (what some teachers have described as "bloodless") teaching-to-the-test moments. Ironically, tests can work well if they are *not* scored, in the form of no-stakes, ungraded quizzes taken at the end of an experience-based learning event. It's better to create a learning moment in which learners actually engage with the topics they are studying and then use score-free tests to prompt memory retrieval!

Education research has also proven that students who believe their learning will enhance their minds, or that it will lead to mastery of a subject, learn more and perform better on tests, even IQ tests, and are more likely to complete courses compared to their score-minded peers.[23] Likewise, students who approach learning with the goal of mastering a subject over their lifetime, as opposed to scoring high in the immediate moment, do better later in life in higher ed and career outcomes.[24] In addition, research shows that people learn best when they direct their own curriculum based on their personal interests—setting priorities, goals, and even assessments themselves.[25] In other words, the learner who looks around themself and determines what will become the window into new knowledge, who learns because they are invested in their own personal growth, is the one who truly succeeds.

Educators have found that even just an hour of growth mindset training can improve students' grades in difficult subjects like math.[26] These principles have also shown to confer benefits during particularly challenging moments in life—such as the transition to middle school or high school. Students who were taught to develop a growth mindset performed better through the course of their new education.[27] Low-performing and at-risk youth have been particularly helped by adopting a growth mindset, leading many education systems to adopt Dweck's and other researchers' teacher and student training programs.[28,29]

In the realm of business, a large body of human resources (HR) research has similarly proven that training leaders in a growth mindset leads to better leadership outcomes, including enhanced leadership awareness, effective leadership strategies, and productive leadership action.[30] Research into employee effectiveness and engagement also has shown upward trends in workplace commitment and satisfaction as well as career outcomes as a result of growth mindset training. There are now a number of corporate programs dedicated to growth coaching, organizational citizenship, and work engagement. Large-scale companies like Microsoft have been training employers, managers, and employees using growth mindset technologies for almost a decade, demonstrating just how an entire company culture can be shifted to embrace growth, and how that culture rooted in growth can translate to overall market success.[31]

Your Brain on Growth

Neuroscientific research into the opposing mentalities of score versus growth also shows us just how having a fluid sense of intelligence can promote neuroplasticity in the brain by prompting the brain to think in new ways and by reducing stress. Using MRIs, electroencephalograms, and other brain scans, neuroscientists have shown that the minds of those who believe in their own neuroplasticity, their own learning potential, and their eventual mastery of knowledge undergo cascades of electrochemical connections when they encounter a challenge, flexing a range of brain regions while exercising and enhancing attention to environmental stimuli.[32] Their brain activity is increased, because they are taking full advantage of their environment, remaining in a reflective state of mind where they are actively seeking opportunities to learn and use failure as feedback.

The brain's response to mistakes has been an especially fruitful research area for neuroscientists studying mindsets. Studies show that people who see their intelligence as growing in response to the environment spark greater brain activity when they make mistakes.[33] Instead of the brain simply noting a conflict, a brain on growth additionally logs the conflict as an error (that is, it realizes a mistake was made) and then puts conscious attention to the error in order to begin the process of correcting it.

Strikingly, studies also show that growth-minded people who rewire their brains to correct their mistakes simultaneously formulate the emotional mechanisms to handle them with ease.[34] So, in addition to seeing errors and correcting them with superior accuracy, they "rebound" from mistakes more easily. They feel motivated to try again and have the optimism to do so. And as we know from stress and outcomes research, they clear a path to test better and score higher.

Related to this emotional mettle developed in response to momentary stimuli, growth-minded people also display emotional mettle in response to overt criticism and threats to their competence. One study involving intelligence tests gave participants fake low scores with varying levels of feedback about those scores.[35] Those possessing a fluid sense of intelligence were not only able to bounce back from news of low scores, but they were also able to bounce back from being told that their scores indicated low ability relative to others.[36] Meanwhile, the brain activity of those with a fixed sense of intelligence experienced low scores as failure and detailed feedback as harsh criticism. Their brains even showed evidence that they had set in motion a debilitating stress cycle that mirrors how brains usually experience punishment. In other words, they punished themselves instead of deflecting and leveraging feedback to their advantage.

Research in children has only echoed these findings for adults, which is a red flag for any parents or caregivers out there who might be thinking (and modeling) a score mindset or a fixed sense of intelligence.[37] Children's neural networks are developing connections that will serve them throughout their lives. Will they punish themselves and learn to fear learning? Or will they use mistakes to enhance their perceptiveness and develop a lifelong love of learning?

In biological terms, there is another value of challenging yourself to rethink your intelligence and your journey: reducing your stress levels. Attuning your brain to the iterative and generative nature of life, and the ever-present potential of learning and improving your mastery of knowledge, helps your brain respond to the environment with positivity and hopefulness. In high-stress moments, it's easy to fall into an all-or-nothing mentality and dread the consequences of making a mistake. But our inherent neuroplasticity tells us that life is not a one-shot goal. Implementing a fluid, processual sense of intelligence allows you to align with the journey of learning while removing the pressures of the endgame.

Stress research has shown that students who saw intelligence as fixed had higher cortisol levels when their grades were declining, and they had sustained high cortisol levels when day-to-day academic stressors came their way.[38] Meanwhile, students who saw their intelligence as fluid not only coped better with academic stress, but were also able to let go of stressors immediately, thereby reducing cortisol levels and ensuring proper HPA-axis functioning.[39] Other studies have demonstrated that growth mindsets predict higher psychological well-being and school and work engagement.[40]

Studies focused on setbacks have shown that those with a growth mindset cope better with setbacks and are motivated to achieve better.[41] Stress research in workplace settings has also shown that a growth mindset not only mediates job stress but even reduces the amount of "counterproductive workplace behavior" in an organization, such as aggression, harassment, and absenteeism.[42]

Most of the research proving the ways that a growth mindset offsets stress examines people living in situations that are relatively normal to them. But since the onset of the COVID pandemic in 2019, researchers have also studied how a growth mindset can help during exceptionally stressful times. Studies have shown that focusing on the mind's flexibility and one's unending supply of intelligence, creativity, and ingenuity has rescued many from the psychological distress and post-traumatic stress of essential work, loss and grieving, and even loneliness.[43] Healthcare providers and biomedical scientists have found that adopting this way of thinking can reduce depression, substance abuse, and self-injury and increase healthy habits such as quality sleeping, eating, and exercise. Even substance abuse programs initiated in the pandemic have shown better outcomes when patients in recovery have adopted a fluid notion of their own intelligence and begun seeing their recovery in terms of a lifelong practice.[44]

This research shows that, without a doubt, growth mindsets allow our brains to flourish and at the same time they lower our stress levels. And, as we have seen, reducing stress is key to maintaining a healthy epigenome. So, in addition to giving you a more realistic picture of what's really going on with your brain and your intelligence, seeing your potential for growth allows your body to turn on the genes that are beneficial to you. As Dweck says, every moment that you challenge yourself, your brain makes stronger connections within itself. Seizing the learning moment takes this growth beyond the brain to your wider physiological systems, empowering you to live out your true potential.

CHAPTER 6

From Mind to Mindful

Use your life to wake you up.

—Pema Chödrön

Until now, our exploration of intelligence has been largely focused on brain function and fluidity—the traditional markers of "smarts." But, of course, in our new, more expansive framework of intelligence, the *whole* of our mental health must be considered. And so, let's consider some of the other factors that contribute to a high-functioning brain, such as focus and a sense of well-being.

One of the most important dimensions to our brain function is our ability to be attuned to what is happening around as the moment unfolds, a skill that can be honed through cultivating mindfulness. Mindfulness is typically defined as a practice in which you place your attention on what is happening in the moment without judgment.[1] Instead of letting your mind wander to the past, or anticipating what's yet to come, you focus on fully participating in the present. Some people achieve mindfulness though meditation or yoga, which can be very helpful tools, but directing your attention to the present moment does not require that you adopt an Eastern philosophy or sit still away from the world.[2] Mindfulness in the context of expanding intelligence is about the everyday interactions you are having with your present-moment environment and the resources you allow yourself to take advantage of as you move through your day. It is a process of honing your awareness no matter what you are doing.

One of the challenges we face in fully tuning into our environment is that our brains tend to wander. As anyone who has ever sat through an especially long lecture or slogged through a tough read or endured a boring movie can attest, forcing our brain to focus on what we want it to focus on can sometimes feel like an impossible challenge. We may find ourselves bothered by the background noise around us—the sound of paper crinkling or someone coughing—or even consumed by physical distractions—the sudden ache in our back, the tiredness in our eyes.

Another challenge to our focus is the emotional noise that stress creates. As we've learned, stress hurts us on every level, from our physical health to our psychological well-being to our brain function and learning. Stress impedes clear thinking and it worsens our performance of mental tasks. We all know how this feels, too. Often when the stakes are highest—and we are the most stressed out—it can feel impossible to focus on the present moment. Instead we may ruminate on a recent mistake, or daydream about a better future. Eventually our frustration with being unable to focus on the task at hand creates more stress, and begets less focus, and so on. Stress can push us into what feels like a hopeless spiral of anxiety and distraction. This is when mindfulness can come to the rescue.

Mindfulness acts like a homing device that zeros in our attention on the most important elements of our environment and quiets the distractions, both external and internal. It can also help us reduce our stress levels significantly.[3] When practiced over time, it gradually instills a more stable state of mind, one that allows for thoughtful reflection rather than impulsive reaction, and in which your emotions can be marshaled to *empower* your cognition.[4]

In helping us to focus our attention on where we want to put it, mindfulness facilitates the acquisition of new knowledge and teaches us a better way of perceiving the world. When the mind can become still, we can harmonize our emotions and thoughts while also allowing our learning brain to harmonize with our external environment. As Zen master Pema Chödrön has suggested, mindfulness truly wakes us up.

Your Brain on Mindfulness

Just as with growth thinking, your brainpower is enhanced by mindful thinking. Brain imaging studies have revealed that practicing mindfulness stimulates the parts of the brain that govern focus and learning.[5] Normally, our minds are set to wandering. We scan our environments superficially, bouncing from one bit of information to the next like a bee going from flower to flower as it pollinates. When engaging in mindfulness practice, your brain takes you out of your default mind-wandering state, which is activated in the posterior regions of the cingulate cortex, to instead signal your anterior regions that a distraction is present.[6] Those anterior parts then flag your prefrontal cortex (remember that part of the brain that calms you down when your amygdala cries "stressor!"), which works with other regions of your brain to reorient your attention to more important information.[7] The brain continues this noise-canceling, attention-honing workout, becoming more focused with every mindful thought.

Indeed, there is a large and ever-growing body of neuroscientific research that suggests mindfulness enhances the ability to quickly react to stimuli in your environment,[8] to sustain attention to visual and auditory input like images and sounds, to interpret that information, and to reduce what neuroscientists call "attentional blink"[9]—your mind's ability to identify two visual stimuli in quick succession. Brain imaging has also revealed that mindfulness practice stimulates the parts of your brain that ensure you are optimally syncing your attention with your visuo-spatial processing, working memory, and the suite of functions that help you plan your next moves (what researchers call "executive functioning").[10]

Neuroscience has also demonstrated a positive correlation

between mindfulness, neurogenesis (new brain cell growth), and intellectual performance.[11] Brain scans of mindfulness trainees have shown substantial growth of brain matter in the hippocampus, prefrontal cortex, and other regions used in complex thinking and problem-solving.[12]

We can see the effects of mindfulness on the brain in terms of speed and quality of completing cognitive tasks. Researchers have conducted randomized control trials where groups of participants are randomly assigned a mindfulness practice or not, and then the participants are scored on a range of cognitive tests.[13] In fact, so many of these studies have now been conducted that we are able to perform what scientists call "meta-analyses" of the entire body of research to pinpoint the most definitive findings.[14] These neuroscientific analyses show us that mindful brains consistently outperform nonmindful brains in attention, memory, executive function, and high-order functions like language processing and abstract thought (and this performance has showed sustained improvement even in programs lasting several minutes a day for less than a month!).[15] Also, older study participants who maintain a mindfulness practice have displayed improved cognitive functions on par with the effects of some dementia medication, indicating that mindfulness also has neuroprotective benefits and may offer a buffer against age-related cognitive decline.[16] Several review analyses have found that mindfulness practice can thus be a safe complement or alternative to pharmaceuticals, especially in the case of dosage-vulnerable populations like children and the advanced elderly.[17]

In terms of the genetic factors we need to support our intelligence, genetics has also demonstrated an important link between mindfulness practice, neuroplasticity, and epigenetics. Genetic researchers have witnessed the power of mindfulness practice to kick-start neurogenesis by allowing our brains to switch on our "good" genes to begin creating new cells. In fact, multiple studies have found that mindfulness interventions can increase the expression of genes beneficial to brain cell production and signaling, meanwhile decreasing the expression of harmful methylators that prohibit it.[18] In other words, mindfulness practice can help our brains express our genes in ways that support our neuroplasticity. Studies have seen this kind of enhanced epigenetic function and brain cell function in patients with advanced neurological disorders, such as Alzheimer's, major depressive disorder, and PTSD, as well as in healthy subjects who are at risk for these disorders.[19] These epigenetic findings correspond to the brain imaging research just discussed that suggests that mindfulness practice can have disease-preventing, neuroprotective benefits for everyone.[20]

The sum of this research shows us that mindfulness practice can spur the growth of new brain cells and boost the brain's capacity to rewire in ways that support our cognitive abilities and our epigenetic health—including slowing the brain's aging process. Also, a number of these studies have traced how these neurological improvements are passed on via neuroprotective epigenetic sequences. That means it is worthwhile to think of the lasting benefits of mindfulness not only in terms of your cognitive ability now, but also in terms of your future and the future of further generations.

Taking the Stress Out of Life

One of the most touted benefits of mindfulness is stress reduction—and as we know, stress is one of the most potent threats to our cognitive (and overall) well-being.[21] In our earlier overview of the stress cycle, we discussed the ways in which the brain marshals various resources and regions in response to a stressor. The brain then signals the body to flood itself with cortisol so as to react to the environment and begin to problem-solve. A healthy mind keeps its cortisol levels elevated only long enough to confront the stressor, decide how to proceed, and take restorative action. Cortisol must abate so that the brain can begin to perceive the stressor calmly and rationally, and to construct a rational plan of action.

Brain imaging has revealed that mindfulness engages these same parts of your brain, helping you to stay in a calm, rational state of mind even when your body is flooded with stress hormones.[22] In fact, the HPA axis, which is controlled by the hippocampus region of the brain, has been shown to grow more neurons from routine mindfulness practice.[23] It literally bulks up the areas of the brain that tell your mind and body that you are perceiving an intellectual challenge and not a mortal threat. Brain imaging has likewise revealed that mindfulness training can reduce levels of cytokines, those tiny proteins that your immune system releases to attack intruders, but which can cause damaging system-wide inflammation in the process (especially if your body is misperceiving a threat, as in the case of autoimmune disease).[24] This research suggests that mindfulness reorients your stress reaction throughout your whole body, preventing it from going into "fight or flight" mode when, instead, it can steadily handle, learn, and grow from what you are experiencing.

Indeed, there is an evidence-based stress-reduction program called "mindfulness-based stress reduction" (MBSR) that uses mindfulness practice for this sole objective: to reduce the experience of stress and minimize its downstream effects.[25] This practice grew out of the work of molecular biologist Jon Kabat-Zinn, a scientist analyzing human genetics, disease, and immunology who established the first stress-reduction clinic using the insights of genetics and neuroscience.[26] Housed in the University of Massachusetts, Kabat-Zinn's lab initially proved that meditation and yoga attuned to moment-to-moment awareness helped patients and healthcare providers reduce their stress levels in medical settings.[27] Yet it didn't take long for Kabat-Zinn and other scientists to recognize the power of mindfulness practice for people living and working outside of a clinical environment.[28] Since the 1980s, researchers around the world have set up their own MBSR labs to devise ways to tailor and improve mindfulness practice beyond meditation and yoga, and for their own diverse constituencies and communities.

Across these studies, MBSR interventions have consistently proven to be particularly beneficial in reducing stress for those who have been diagnosed with anxiety disorders and mood disorders, and those with intermittent stress disorders, such as panic attacks.[29] Such studies have also revealed that even small investments of mindfulness practice yield sustained benefits.[30] Across the board, time-limited interventions such as a onetime two-week meditation program or a four-day breathing workshop have been shown to impact practitioners for months and years to come, improving their mental health and at times reversing the course of a disease.[31] Mindfulness has contributed to people feeling cured from devastating illnesses like major depression and generalized anxiety disorder, among others.[32]

In addition to improving mental health outcomes, MBSR has shown that mindfulness practice helps patients suffering from other diseases, such as cancer or cardiovascular disease, to experience less fear about their circumstances and feel more in control of their lives.[33] Studies with patients who have received terminal diagnoses or otherwise poor prognoses have suggested that mindfulness can improve brain health while bettering disease outcomes by way of sweeping mental and physiological improvements.[34] Mindfulness practice has been shown to reduce muscle tension,[35] increase oxygen flow to the brain,[36] and strengthen neurological functioning.[37] And improvement of these processes at the basis of a person's wellness has proven to feed back into a patient's disease progression and overall health.[38] Studies have even shown mindfulness practice to bring these mind-body improvements to healthy partners and caregivers of diseased patients, illuminating the power of mindfulness to heal by way of improving the nonclinical elements and relationships in patients' lives.[39] And while the body of MBSR research is vast, these studies focused on the mental-emotional, mind-body gains of terminally ill patients highlight how even the most stressed of us—people who are literally fighting for their lives—can reap powerful benefits from practicing mindfulness.

Putting Your Mind to Good Use

We know that mindfulness can be used to enhance our ability to focus, tune in to the learning opportunities within our environment, help reduce our stress levels, and allow us to make better, more rational decisions. But mindfulness also has a "softer" benefit, one that's related to emotional growth. The same stress reduction that allows you to slow down and think clearly also helps to promote a greater sense of well-being and feelings of happiness.[40] In turn, this emotional awakening feeds back into your ability to think clearly and cogently, which improves your ability to perform mental tasks. It's a feedback loop with compounding benefits.

Cognitive psychologists have found that mindfulness enables research subjects to increase their self-awareness and resilience.[41] These studies demonstrate that those who regularly practice mindfulness gain perspective on their thoughts and feelings, and are better able to accept the ups and downs of life, seeing them as temporary challenges. With a less judgmental outlook, subjects report more positive feelings, greater acceptance of their negative feelings, and quicker bounce back from negative reactions. Stress is reduced, while positivity and curiosity are increased, feeding back into their ability to focus their attention and achieve their goals.

Education research has also documented a mindfulness practice's beneficial effects on emotions and performance in academic settings. In schools, mindfulness programs have been proven to help students regulate their emotions[42] and increase self-compassion,[43] instilling stress-reducing thought cycles that free up their minds to perform better. Case-control studies, where one classroom or school has been given an intervention and another has not, have shown that treatment students were markedly less stressed, more focused, and more emotionally regulated than control group students.[44] They were better able to learn new material and to practice self-control as they were doing it. In one Harvard–Massachusetts Institute of Technology study conducted on sixth graders in Boston, for example, students who took mindfulness training in place of computer coding were better able to control their emotions.[45] Further brain scans of these students confirmed that their amygdalas—the brain's fear center—were less sensitive to negative visual stimuli. In short, these kids' brains thrived on mindfulness. They experienced less stress, better focus, and reduced emotional reactivity—all foundational qualities to facilitate learning.

Mindfulness research in educational settings has also found that trainees display more cooperative social behaviors[46] and resilience,[47] and less hostility and violent behavior[48] toward others. These findings have proven particularly salient in learning environments where schools are impoverished and where students are routinely exposed to stressful life events like violence in their neighborhoods, loss of loved ones, or experiencing instability in housing. In one resource-poor elementary school in Nashville, Tennessee, where 90 percent of the students live below the poverty line, behavioral referrals like sending students to the principal's office went down by 80 percent following one mindfulness intervention.[49] Students in neglected environments like these have demonstrated enhanced executive functioning, which controls our judgment and stress response, as well as improved attention and focus, which is essential to academic and social learning, meanwhile also showing greater self-awareness, self-regulation, and relationship skills. And students in the direst of circumstances, who are experiencing PTSD from living in devastating environments, have shown marked improvements in their disease severity. As one principal put it, "We've given them avenues to work on mindfulness, to work on just calming themselves, getting to their center place where they can just be children." Meanwhile, grades, test scores, academic performance, and attendance have improved across the board for kids who have had some form of mindfulness-based education.

It will come as no surprise that cognitive research in work settings has only mirrored these improvements that we see in educational settings, proving the truism in business administration and economics that organizational health depends on employee health. Mindfulness programs there have similarly been proven to boost comprehension and rational thinking while promoting wellness, making work more productive and less stressful.[50] Mindfulness has also increased work engagement, productivity, and feelings of job satisfaction, while boosting levels of job performance.[51] Workplaces see greater levels of trust and compassion, while workers see greater levels of pride in goal attainment and sense of personal growth.[52] Some studies have even found mindfulness programs to simultaneously improve employees' feelings of commitment to their work and their work-life balance.[53] In other words, they have had a spillover effect at home, promoting productivity, satisfaction, and well-being at both places at the same time.

Again, these improvements have proven especially salient to wanting work environments, such as factories in which workers engage in monotonous work. In one study of blue-collar workers at a manufacturing plant in Mexico, researchers found that mindfulness training made workers feel less bored, more engaged in their work, and more satisfied.[54] And trainees' boost in feelings of job satisfaction led to a marked increase in the quality of their work. Mindfulness training even decreased a worker's inclination to want to quit their job, which helped lower the levels of attrition that were afflicting the plant.

Indeed, an altered organizational culture—one that prioritizes mindfulness for most if not all participants—has proven to be invaluable in bringing about individual emotional growth and organizational improvement. Indeed, by 2015 more than 80 percent of medical schools had their own mindfulness programs.[55] In a nationally representative survey in the United States conducted in 2018, researchers found that approximately 60 percent of mid- to large-sized companies had in-house mindfulness offerings.[56] Just about every industry out there, from furniture retail and clothing manufacture to finance and food, has developed its own programs.

Still, it is Big Tech and healthcare companies that have led the pack with their emphasis on specialized apps, online and in-person trainings, and company software packages that all members can use at no cost. There, a mindfulness-infused work culture is the goal of many executives, and increasingly it is, as SAP's mindfulness director Peter Bostelmann calls it, the "new normal," because they find that it makes good business sense. Bostelmann, for one, reports that more than 6,000 of SAP's employees have taken their mindfulness courses, which has spurred significant increases in job engagement and employee trust of management.[57] Health insurance company Aetna, which offers yoga and meditation to all of its 50,000 employees and has trained at least 13,000 employees in its signature mindfulness courses, estimates that it has generated $3,000 per employee thanks to increased worker well-being.[58] Aetna's CEO, Mark Bertolini, who has been at the head of the company's in-house rollout and who encourages meditation in executive team meetings, has found that employee stress levels have decreased approximately 28 percent since mindfulness training hit the scene in his organization.[59]

While research on the mind-body benefits and performance improvements of mindfulness practices has largely focused on education and work sites, it is important to note that these benefits have been observed in other environments thanks to an abundance of free community mindfulness programs. Many communities provide programs to the public at healthcare, government, and academic institutions that, in addition to cognitive and emotional improvement, focus on teaching compassion and care for others. My home community of Princeton, New Jersey, for instance, offers free six-week mindful thinking programs for kids, teens, or adults at various community parks and recreation sites, while the university offers a public archive of guided meditations and virtual stress-reduction courses. Hospital-based programs like Johns Hopkins All Children's Hospital in St. Petersburg, Florida, and the Children's Hospital of Philadelphia Mindfulness Program bring kids who are inpatients together to learn how to center themselves and interact with each other and the hospital community more peacefully and compassionately. Other facility-based programs like the Mindfulness Minutes series at West Virginia University's Center for Excellence in Disabilities similarly teach mindfulness so that kids with intellectual or developmental disabilities can boost concentration, relaxation, and social-emotional learning so that they can have healthier social interactions.

There are also a number of international online communities where diverse groups can train and grow together thanks to the open-education platforms like FutureLearn, Khan Academy, and University of the People. These mindfulness spaces, which crowdsource expertise from clinicians and academics from all over the world, provide free mindfulness education to all anywhere, anytime. Some also provide links to free meditation and sound therapy apps that can be downloaded and adapted to other community contexts as well as virtual gatherings.

Finally, there are many mindfulness communities developed by and for groups of people that focus on emotional growth through healing the collective spirit and promoting social equality. Offerings from sex and gender centers like the Women's Meditation Network or the LGBTQ Tuned In! program help people develop their own mindfulness practice while advancing a greater humanistic agenda of communal compassion. All over the world there are also ethnic and racial minority advocacy programs developed for what those of us in the United States presently refer to as Black, Indigenous, and People of Color, or BIPOC. The Mindfulness Self-Compassion for BIPOC, for example, focuses on "self-compassion in community" to strengthen the inner self while enriching and fighting for the outer world. These mindfulness engagements foster reflexive conversation among community members about compassion-building while encouraging trainees to become more self- and socially aware.

Becoming Mindful

Given the overwhelming evidence of the benefits of mindfulness practices on cognitive health, emotional growth, and performance—and its relative accessibility to all—I've made mindfulness one of the core pillars of my intelligence paradigm. I want to encourage anyone interested in improving their mental performance to embed mindfulness into their lives.

The question then becomes—how do we do this in a practical way? The key is to approach your every interaction as an opportunity to become more present and attuned to your environment. This means starting at home base. How can you slow down, despite the rush to work or to care for others? How can you learn to return to stillness, to alertness, to openness to what the moment is providing you in your very own sanctuary?

I use three strategies at home: centering, breathing, and meditation. Focusing on the here and now is like flexing a muscle. The more you flex it, the more capable you are of maximizing your focus. So, when I begin a mundane task like cleaning the dishes or putting in the laundry, I practice centering myself in the moment. I activate my senses of sight, hearing, and touch so that I can better think about what I am doing and learn from my present experience. I take a task that would otherwise be mindless and make it into my personal mindfulness workout. This has helped me to see other opportunities in my day-to-day for centering and it has improved my overall ability to be present in what I do.

Centering is very important to me because breathing and meditation require downtime. But breathing and meditation are also important to me because they provide a more quieting form of mindfulness within which my brain can practice focusing. Luckily you don't have to reinvent the wheel. There are many proven practices and modalities to choose from, such as box breathing (inhale for four seconds, hold your breath for four, and exhale for four) or Zen meditation (count your breaths from 1 to 10 and repeat). I prefer guided audio body scan meditations in which a guide helps me to focus on the various parts of my body as a way of relaxing into a more attentive mode of thinking. Today there are plenty of apps and online audio files to download, use, and return to whenever you have time. And as we have seen, there are nonprofit mindfulness centers across the country (and the world) that also offer complimentary workshops and classes and bring together diverse groups of people to practice and discuss breathing and meditation interventions.

In the workplace, we can follow the lead of those Big Tech companies that have invested a great deal to change their work cultures. One of the earliest and most prolific examples is Google's Search Inside Yourself program, which one Googler initiated in 2007.[60] What started out as an internal series of meditation and compassion courses and lectures metamorphosed into an external stand-alone mindfulness institute that today services more than fifty countries around the world. The central program now consists of a one-month interactive online course that includes experiential and writing exercises, group work, and attention-training. The institute also has a number of spin-off programs that combine compassion-training with resilience-building and other forms of emotional and mental strengthening, so that workers everywhere and in any profession can boost their centeredness and focus. In place of the specifics of Search Inside Yourself, you can use their model as a template, adapting several minutes of your break time each day to engage in one or another of their practices.

Another type of work-based program that you can adopt is the "employee mindfulness challenge," exemplified by Apple's Mindfulness Challenge. In 2018, Apple began partnering with journalist Dan Harris's Ten Percent Happier program to provide an app-based course in which employees had to meditate for a certain amount of time every day for twenty-five days.[61] Social media companies like LinkedIn have also adopted the challenge method, incentivizing employees to literally work meditation into their daily lives, practicing at work every single day.[62] You can engage in your own Mindfulness at Work challenge where you commit five minutes of your time each day to meditating. Once you have built up that mindfulness muscle, you can increase your time or frequency.

Many companies have developed mindfulness spaces to encourage employees to develop a daily mindfulness practice. Steve Jobs is famous for introducing the Zen Room to Apple's headquarters in Cupertino, California, where he commanded employees to sit for thirty minutes a day. Now companies like HBO, Pearson, Nike, and Yahoo all have their own versions of mindfulness spaces. Salesforce CEO Marc Benioff is such a fan that he has dubbed his company's spaces "mindfulness zones," where employees are invited to retreat to at any point in the day.[63] Benioff has had zones placed on every floor of over half of Salesforce's offices in hopes that employees will equate its entrepreneurial culture with one of mindfulness. Though an employer-provided mindfulness zone may be out of reach for most of us, demarcating one area of your workplace as a mindfulness space can help you to build your practice. I have worked as a gas station pump attendant, a pizza deliverer, a teacher, and a writer, so there are many "stations" at which I could have designated a centering or breathing zone such as the squeegee stand, the front seat of my car, the lectern, or my office chair. Today, as my remote work has increased and I find myself working wherever I can post up with my computer, I have committed to beginning any work session with a five-minute seated centering exercise.

My experiences have taught me that there are many ways to take advantage of my environment and transform it into a brain-strengthening workout. Mindfulness helps with the immediate goal of leveraging any place to help hone the attention as well as the end goal of leveraging every moment to maximize the mind.

CHAPTER 7

Learning to Connect

Individually we are one drop. Together we are an ocean.

—Ryūnosuke Akutagawa

We have seen how our intelligence is supported when we adopt a growth mindset and practice mindfulness. These strategies shift our notion of intelligence from quantity (high rank/high score) to quality (seeing our innate potential, relishing in the experience of learning). They also support our inherent neuroplasticity and our mental and emotional health. The final core pillar of my intelligence paradigm builds on these strategies to improve the quality of our experience of learning. It teaches us to go from learning-to-score to learning-to-connect.

As an intellectual lifer and imparter, I have experienced firsthand what studies time and again have shown: learning works better when ideas are connected to a context, to action, and to others. This is true whether we are talking about learning a new skill or concept, or solving problems and making decisions. When learning with others, our brains are motivated and driven to the best of our potential.

So, when seizing that learning moment, it is important to seek out connection and to encourage communion in thought. You don't have to go it alone. In fact, as we will soon see, the research clearly shows that this is a terrible strategy. Instead, think of yourself as an intrepid explorer and your loved ones and colleagues as your fellow crew. You are on this journey of life together and it is the collaborative learning that will be most generative for you.

Lucid Learning

The human mind is hardwired to constantly learn. "Life is learning" may ring true in an abstract sense, but it is decidedly true from a biological standpoint. We spend nearly our entire lives in what neuroscientists call a "default" state of mind. We are constantly scanning our environment, looking for new information about our world, and comparing that information to what we know while making decisions about how to proceed. This, in a nutshell, is using our intelligence.

Still, how much of it is lucid? How much is self-aware? And how much do we know about learning itself?

Even academics like myself generally spend very little time thinking about learning. What different kinds of styles are there? What is most effective for the general population?

What works best for different kinds of people with different abilities and different resources? Luckily for us, there is a prolific science of learning that answers these questions and even compares them all to offer us a road map for understanding which learning styles are most naturally attuned to our human minds and bodies.

Education research shows that there are two main approaches to learning. First, there are the traditional forms of learning and instruction that people now refer to as "teacher-centered." These approaches assume that learning happens in an educational setting, where teachers and their books hold all the answers.[1] These approaches see the moment-to-moment experience as unhelpful to, or even at odds with, learning. Under this paradigm, for example, lecturing students about the color spectrum using a textbook would be superior to having students work together to create prisms of light themselves.

This model is built on the presumption that learning is vertical, with knowledge being transmitted from one individual to another. A person who possesses knowledge (usually an older, accredited expert of some kind) communicates to one who doesn't (the pupil). The instructor does the doing, lecturing, or reading aloud, while the learner passively accepts—and is meant to integrate and retain—the information offered to them.

Learning that emphasizes collaborative action, social interaction, and emotional and social learning is at the heart of the early childhood education we strive to offer toddlers.

However, teacher-centered learning quickly replaces this approach when grading and testing come into play in elementary school.[2] Then, classrooms are arranged to facilitate the vertical transmission of knowledge, with students seated in a face-forward grid that encourages them to ignore all else but the teacher. This setup may trigger memories from your own years of schooling, whether primary or secondary. Most formal education systems have embraced this traditional style of learning for centuries and have been thoroughly edified by it.

Alternatively, there are what researchers refer to as "progressive" forms of learning, forms that put learners and their collaborative hands-on experiences at the center of learning. These "learner-centered" approaches pivot on teamwork so that learning is cooperative, inquisitive, problem-based, synergistic, and meant to reflect life as it is always experienced, from moment to moment.[3]

In formal educational settings, teachers who embrace progressive learning act as facilitators of knowledge acquisition. They simply encourage students to ask big questions about a given subject.[4] They may prompt students with specific articles or videos. Or they may ask them to independently research and collect evidence from their own homes and communities and come to class ready to share what they've discovered.[5] Under this banner, learning is seen as self-motivated from a person's innate curiosity. It is also seen as inherently communal in that it is sparked by our basic need to share knowledge and improve our environment for the benefit of all.

Here, there is no need for desks or lecterns because learning happens when all are thinking together, discussing together. And the learning doesn't start or stop in that space. Rather, it is unending and embedded in the broader context of life, true to the meaning of life as learning.

Connected Learning

One collaborative learning style that is exemplary of this latter mode is what experts call "connected learning." Connected learning asks a student to learn new concepts and skills by creating a project with others.[6] It encourages social connection, with an emphasis on utilizing a student's personal interests and experiences in the real world to grasp new ideas and to make critical decisions for the collective future.[7]

Inherent within connected learning is a reciprocal dimension that mirrors real knowledge acquisition and decision-making outside the educational setting. Learners depend on one another to generate intelligent questions about and intelligent solutions to a problem. They learn together and their success depends on one another. They are not graded against or in competition with one another. They may be responsible for their own piece of a bigger project, but they are accountable to each other in ensuring its overall outcome.

Most connected learning projects include a goal of civic engagement, thereby tying the students' achievement to their classmates' progress as well as real social progress.[8] Many projects also have a digital dimension, encouraging student learners to tap all the resources they can outside the classroom at every step of the way.[9] Learners are encouraged to connect to the world via the web and via remote communications, even if they are problem-solving in person. In this model, technology is viewed as a learner's friend, a means of being and staying socially and globally networked.

For example, let's say you want to impart the basic mathematical skill of counting and subtracting numbers to a group of toddlers, as I do with my three toddlers at home. You can go the traditional route where you explain what the numbers are and then write them down on a piece of paper, spelling out "2 – 1 = ?," and hand them to each to decipher and solve. You can even represent the amounts with lines, writing "II – I" and asking them each on their own to point and count.

But see what happens when you opt for progressive, when you go collaborative, and when you get connected. Try having the little ones set up a dollar store where they each have a couple of dollars to buy a present for the other. When Toni uses one of her two dollars to "buy" a crayon for Amir, something different happens. Not only is the learning happening in a real-world experience that they want to be in, but they are using their brains to literally connect the dots via their personal connection to each other. They are learning to connect to information and new knowledge just as they learn to be a stand-up sibling or friend.

In adult education, connected learning starts by asking a group of student learners to share stories about a specific matter while they are virtually connected to or at the site where the learning is taking place. The group generates questions together, poses a problem to solve together, and gets down to the business of solving it collectively.

For example, in my undergraduate and graduate courses, I teach the concept of health disparities. Instead of relying on textbooks and whiteboards, I begin by having my students collectively brainstorm a list of the top killer diseases that they've have heard are more prevalent in one national subpopulation than another. They first respond with rarer deadly diseases like sickle cell anemia, which they associate with Black populations, and cystic fibrosis, which they associate with White populations. But then they move on to common killers like hypertension and kidney failure, which they associate with Black populations, and Alzheimer's and stroke, which they associate with White populations. Using their assumptions about racial disease prevalence as a hypothesis, we then review the actual epidemiological statistics on the diseases they've identified. I bring in my expertise on the genetics of the disorders to guide them toward queries that can show them that, genetically speaking, diseases travel along narrow ethnic lines and not broad racial ones. I then have my students conduct research on the inequities in health care that people of different racial backgrounds face. Through this exercise, we together learn about the social underpinnings of common disease disparities, like those pertaining to stroke and hypertension, which may appear to be a function of genetics but are, in fact, socially correctable.

A key component of connected learning is encouraging

learners to share their personal connections to the material. We follow our initial sharing exercise by working together in teams to flash-research all we can find on three or four cases, and visiting public health websites like CDC.gov to learn how our public health administration is characterizing and battling these diseases, as well as scholarly databases like PubMed to read abstracts on social epidemiological trends.

Collaborative action outside the traditional learning facility is another core part of connected learning.[10] If we are at a health sciences campus, we might talk to healthcare providers and researchers who work with patients living with the diseases we're considering. Instead of listening to their professor tell them what to know or simply reading about diseases in a textbook, my students are able to empathically and reciprocally get the reasoning going, that qualitative wheel turning, and that reflective mind working. Because real people in the classroom have real connections to these diseases, and real people outside the classroom are working on these very problems often just steps away. And even if you have no relation to their prevalence yourself, you're suddenly only one degree of separation away from a more intimate and urgent understanding of the subject because you have partnered, queried, and problem-solved with people who have.

In the workplace, we can instill connected learning within our teams in order to foster productivity and creativity. Personally, I approach work tasks as collaborative intellectual opportunities in which the social effects of the outcome can be felt. As I manage, delegate, and work with other team leaders in my virtual research lab, I look for ways to make problem-solving communal and decision-making equitable.

I have also removed the silos that traditionally structure my workplace in an attempt to reduce job hierarchies and open up the office spaces that typically keep me physically apart from others. During our two years of working remotely, for example, I used my personal Zoom room as an open-plan research lab. Before that, I held in-person team meetings around a shared meal. Sharing the research experience in real time has made my teams closer and our work stronger. Together, we have managed to obtain large-scale research funding, execute and complete national and international studies, prepare manuscripts for publication, and see our numerous books and articles hit the stands.

So, no matter how physically proximal you are able to be, you can attempt to generate closeness and community in the workplace so that everyone is able to see how they benefit from the work they are doing. You can also try to achieve tasks in teams as much as possible. Not only will this generate buy-in from those working with you, but it will generate belonging, which is one of the key elements in a happy, stress-free, and productive work environment.

Your Brain on Connection

Like growth-oriented and mindful thinking, progressive and reciprocal learning is better for your brain. Neuroscience shows us that more and better input equals more and better output: more brain stimulation tapping into a wider array of your intelligence leads to more brain activity, neural "cross-talk" between brain regions, long-term memorization, and understanding.[11] It also means increased satisfaction with the learning process and more confidence in your abilities, which can strengthen your commitment to learning.

Comparisons of brain scans of those engaged in traditional, "passive" learning and progressive, "active" learning show that active, and especially *inter*active, learning stimulates the brain to produce more neural connections across a wider range of brain regions.[12] It also excites the "higher thinking" parts of the brain in your neocortex that don't just file memories to be lost and forgotten, but rather allow you to analyze, evaluate, and create. The brain's ability to memorize information also improves. One MRI study showed that increases in brain activity contributing to a learner's improved memorization skills could be observed days following an intervention.[13] The scientists conducting this study concluded that not only does an active learner's performance get a boost, but their brain stores new patterns of connectivity that can be reinvigorated at each future prompting.

Brain imaging also offers evidence that learning that allows the student to lead the process with their own questions and interests increases the brain's processing speed, agility, and ability to correct errors.[14] In a battery of tests, learner-centered, inquisitive learning spurred greater cognitive functionality in brain regions that aid in judgment, decision-making, impulse control, and empathy.

A fruitful new area of research on collaborative learning in action additionally reveals that progressive learning works so well because interaction synergistically stimulates the brains of all involved to produce learning-supportive brain hormones. Neuroscientists performing a methodology called "hyperscanning," in which multiple brains are analyzed simultaneously within the context of the same circumstances, have confirmed that collaborative learning sparks all connected learners to experience feelings of well-being and contentment, and at times even excitement.[15] This synergistic chemical activity incites brain hormone cascades that boost motivation, satisfaction, and focus. They also put a kind of positive emotional stamp on the learning process, such that learners come to associate the learning process with enjoyment, reinforcing a love of learning.

All this connectivity is crucial to your brain's health because of one essential principle of neuroplasticity: your brain is malleable, so how you use it can improve it. By adding the communicative and synergistic dimensions of connected learning, you can stimulate your brain to marshal more of its centers at once. With multiple sensory modalities engaged together, your brain will form new neural networks between the left and right sides. And as new information is stored in multiple sites of the brain, this will beef up the networks within those regions, too. The result will be stronger, improved functionality as well as cellular growth where it counts.

Connection Is Everything

Not only does connected learning foster better brain function; like growth-oriented and mindful thinking, it improves our mental and emotional health. Much like mindfulness—which at its roots emphasizes a sense of interdependence with others—studies have found that progressive and reciprocal learning models reduce stress and anxiety while boosting happiness.[16] Researchers have likewise found that these learning models reduce loneliness while increasing a learner's sense of social support.[17] Some studies have focused on groups that are challenged in unique ways, such as ER doctors[18] who must contend with a highly stressful work environment or international college students[19] who grapple with feelings of isolation. Indeed, research shows that progressive and reciprocal learning are particularly beneficial to groups of learners who are vulnerable to being overwhelmed by the social stress of their environment.[20] Following interventions, study subjects experience less stress and feel more gratified by their social contexts and are therefore able to think more clearly, staying focused on the intellectual tasks put to them.

Genetic research also suggests that the stress reduction of connected learning benefits us through our epigenomes. A wealth of epigenetic studies on social connection has already proved its substantial health benefits via stress reduction, emotional satisfaction, and a better use of our brain cells.[21] Connection inhibits the harmful methylation that turns our "good" genes off, while allowing our brains to exercise better visuo-spatial functioning, and to practice empathic and social learning.[22] While research programs measuring the epigenetic effects of specific learning styles are just getting off the ground, preliminary findings have already shown that progressive approaches can have a protective effect while traditional approaches can do just the opposite.[23] This research has shown that when learning is collaborative, reciprocal, and connected, it can prevent methylation.[24] This research has also shown that learner-centered learning at the earliest stages of life can promote healthy epigenetic patterning in the brain at our most crucial time in neural development.

In short, our epigenetic machinery is setting the right genes to "on," inducing our brains to generate memory chains rooted in real-world events and interactions. Moreover, those extra-empathetic stimuli in the concepts we are discussing, wherein they are contextualized in our lives and our collective fate, help us to synergistically spark our brains to recall memories and respond to our environment rationally. So, while we are utilizing our intelligence, learning new concepts, generating new data, and finding new patterns, we are also programming our epigenomes to spur our genes to produce a healthy brain and mind. The positive epigenetic feedback loop propels us forward, enhancing our future ability to think intelligently.

The Means to a Better End

Research on various models of progressive learning has consistently found that connected learning outperforms the traditional "transmission of knowledge" model. Going as far back as the 1930s we have social scientific evidence that learning that is interactive and cooperative is more likely to help students and educators meet their stated goals.[25] This kind of learning also improves the quality, efficiency, and efficacy of our work. We are literally more productive and produce better results with a collaborative approach.

Fast-forward nearly one hundred years and today we are still finding that progressive learning improves outcomes. In education science, meta-analyses of thousands of studies continue to find that learning outcomes[26] and social-emotional[27] outcomes all benefit from a collaborative approach. These studies show that learning that prioritizes connectivity and reciprocity boosts a person's ability to think critically about new material, to analyze and form unique judgments about data, and to communicate clearly and rationally about the knowledge being generated.

The depth of knowledge that student learners garner from educational tasks is also measurably better in progressive learning models. Students depend less on "surface-level" memorization that too often gets forgotten thanks to our tendency to superficially scan instead of retain information, and more on in-depth analysis that richly engages the various regions of our brain. In addition, students show a marked gain of interpersonal skills and an uptick in self-regulation, dynamics that, as most of us can recall from our school days, can quickly undermine learning in any classroom if not encouraged.[28] Better social and emotional skills lead to more productive and focused educational environments. And for students, learning becomes self-driven and gratifying in emotional as well as intellectual ways.

Workplace research buttresses these findings, proving that a connective approach to working improves productivity and attainment outcomes. We may already know that teamwork is generally a better approach to problem-solving than going at it alone. But workplace research finds that equitable teamwork supported by an intentional connected-learning framework offers even more benefits.[29] Team members listen better to one another, draw on and learn from each person's strengths better, and find their work more stimulating and enjoyable. Their emotional connectivity and intellectual reciprocation lead to buy-in. That buy-in leads to quantitatively and qualitatively better endpoints.

It is perhaps no surprise that connected learning research in work and educational settings has been on the rise as of late. Remote work and school have led many managers and administrators to seek out efficient and efficacious collaborative strategies that work with open technology platforms. Major platforms from companies like IBM and Google have begun to proliferate.

All the while, research on the benefits of connected learning continues to pour in, providing concrete proof that even in remote and hybrid conditions, this type of collaboration improves people's sense of belonging while simultaneously enriching their technical know-how and knowledge outcomes. A connected approach is therefore well suited to a variety of learning situations and circumstances, as well as a range of groups and populations.

Staying Connected

I've incorporated connected learning into my new model for intelligence not only because of the reams of research that offer evidence of its effects, but also because I personally find that I benefit from this model. Much of my work is done individually, but I have found that outside of communal environments, these principles can help reduce stress and jolt me into a better mindset.

When I am feeling overwhelmed by my work, I pause and ask myself three questions: *Am I thinking creatively?* Connected learning rejects rules-based thinking and encourages us to approach our environment with an open mind. This way we can be open to observing new information and coming to our own conclusions about its patterns.

Am I thinking about what interests me? Connected learning encourages us to articulate questions about our problems, within our surroundings, based on our interests. Seeing your environment as yours to probe will help you create that growth and learning mentality and keep it brimming in your every move.

Am I thinking collaboratively? Synergy is real. When you put two or more heads together, you get far more than the sum of their parts. And though not every kind of learning moment will require real-time collaboration, we can all seek out ways to connect our unique tasks to the work of others. Thinking through the work with others' growth in mind also increases our sense of connection and belonging, as well as higher-level thinking, communication, and leadership skills. It helps us gain deeper knowledge of our individual worlds while strengthening our relationships to those within it—even if they are not physically present.

In adding a social and contextualized element to your learning, we can spur our minds to work more filaments of their intellectual muscle and build more gray matter. At the same time, we can trade out engagement that encourages stress-based epigenetic coping for engagement that promotes happiness and healing. The end result will be a richer, more satisfying learning experience that allows us, and our collaborators, to improve cognition and performance, and to do so in ways that support enhanced mental and emotional well-being for all.

PART III

Valuing Intelligence

CHAPTER 8

Getting Smarter as a Society

Our ambitions must be broad enough to
include the aspirations and needs of others for
their sake and our own.

—Cesar Chavez

We've covered a lot of ground in our new intelligence paradigm. We know that everything from our stress levels to our genes to our connection with others impacts our cognitive ability. We know that our environments provide a rich array of learning opportunities every moment of the day if we can learn how to attune to the present moment. We know that our learning environment makes all the difference in how we process, retain, and use information. And we know that to really get "smarter," we must capitalize on the brain's neuroplasticity and build more brain cells and stronger neural networks to support new learning.

All of these factors are influenced by your environment, which is conditioning your ability to optimize your intellect. And when I refer to "environment," I'm talking about the very basics: How is your home life? What is the quality of your relationships? What kind of stimulation are your work, educational, and other social environments providing you? How healthy and safe is your neighborhood and community? Epigenetic science tells us that our genes are turning on and off in response to these wider contextual elements all the time. And as we've seen, epigenetic studies focusing specifically on intelligence find that systemic changes that can benefit our shared environment are needed now.

So, while we have largely focused on how we can cultivate intelligence as individuals, it is clear that considering only ourselves is futile. To transform our internal neural networks, we need to transform our external social networks. Indeed, we need a systems-level approach that addresses collective factors like socioeconomics, education, and nutrition at the community level and beyond. Even national policy shifts within the space of a single generation can have lasting effects on you and yours. Look no further than your societal environment to build healthier networks for all now and all to come.

What's the Weather Looking Like Today?

Before delving into strategies, let's revisit what we know about stress and our personal relationship to the broader environment. Sure, we face our own personal set of stressors that come from our unique home environment, workplace, and location within our familial and community networks. But the many public and organizational settings we live in can also be sources of toxic stress. These more insidious stressors also contribute to a cumulative erosion of our mental and intellectual faculties.

If you recall, "social weathering" describes the wear and tear people incur from living in a stressful world. The concept was first conceived by public health expert Dr. Arline Geronimus when she was conducting her predoctoral and doctoral research at Princeton University and Harvard University.[1] Geronimus found that the stress that Black women faced in their daily lives in the communities in which she studied and worked was leading to worse pregnancy and birth outcomes for them and their babies.[2] Studying Black American women of differing age groups, and longitudinally over time, she and fellow epidemiologists found that wear and tear increased with age. This led them to conclude that stress was having a cumulative effect on the body, prematurely aging those facing routine social, political, and economic adversity.[3]

Today there is a large body of research on social weathering and maternal health effects. However, researchers working in other realms of public health and epidemiology have also found social weathering at play in brain health. Throughout the world, we see mental health disparities affecting people living in poverty, who exist in a constant state of stress and are less likely to have access to mental health services.[4] People of low socioeconomic status, what researchers refer to as "low-income," report higher levels of stress than middle- or high-income people. At the same time, they face constant threats to their financial security, and many report facing threats to their physical safety as well.[5] Low-income individuals have less access to health, educational, and social resources. They also suffer greater exposure to violence and occupational discrimination. As a result, they show higher levels of depression, anxiety, poor sleep, poor health, and suicidal thoughts, and this data holds relatively constant over the course of an individual's life span.

Women, LGBTQ-identified, and gender-nonconforming people of all backgrounds and socioeconomic statuses also report higher stress levels throughout their lifetimes, especially those tasked with caregiving roles. For example, a comparison of financially secure White American mothers caring for a special needs child with financially secure moms who didn't share that responsibility showed that caregiver moms suffered chromosomal damage on par with women much older than them.[6] Their bodies became weathered from the weight of their day-to-day responsibilities despite having ample resources and living in a safe and healthy environment. Similarly, studies on LGBTQ and nonconforming populations have shown that members suffer from high exposures to daily social stressors, such as discrimination, stigma, and bias, which weather their brains and bodies regardless of their socioeconomic status.[7] LGBTQ and nonconforming individuals are also more likely to suffer harassment and abuse, as well as endure traumatic violent events.[8] As a result, they experience higher rates of psychological distress, anxiety, depression, self-harm, and suicidal thoughts.[9] Despite being more likely to access and use mental health services, LGBTQ and nonconforming individuals are still also more likely to be weathered by their environments.

Globally, we correspondingly see the stark neurological effects of social weathering among ethnic and racial minorities such as BIPOC (Black, Indigenous, People of Color) populations. This has led scientists to conclude that daily racism plays a key role in weathering worldwide. Many studies have shown, for example, that racism-based stress leads to more rapid neurodegeneration in patients already suffering from mental illness. In one recent study of the Harvard Aging Brain Study dataset, for example, MRI and PET brain imaging found that older Black Americans with amyloid plaques were suffering decreased cortical thickness compared to White Americans of similar age and brain diagnoses.[10] The stress from their treatment in society had diminished their cortexes, giving them an "older brain age," and putting them at a neurological disadvantage at the onset of neurodegenerative disease.

Studies on healthy individuals have also demonstrated how wear and tear sets in motion downward spirals in the brain and body, and even cognitive dysfunction, early in life. Another study on younger Black Americans exposed to cumulative toxic stress found that they were more likely to experience social isolation and yet less likely to have access to preventative health regimes.[11] They were therefore less likely to engage in preventive health behaviors such as using health screens without being sick. As a result, they suffered from greater inflammation and poorer overall health throughout their life course.

Weathering studies paired with cognitive tests have likewise revealed that experiencing racism-based stress earlier in life leads to neurodegeneration and memory loss over time. Black American youth experiencing discrimination as a result of systemic racism suffer cognitive effects, from decreased memory function to worse test performance.[12] Black American youth experiencing the daily onslaught of racism have additionally shown diminished executive function as compared to non-Hispanic White Americans.[13] Research on Latinx youth has similarly revealed that "stereotype threat," or the fear of confirming a negative stereotype about the social group a person belongs to, reduces working memory and therefore leads to worse performance outcomes in the short term as well as memory loss over the long term.[14] Studies like these highlight the fact that unequal environments precipitate and hasten cognitive decline, regardless of a person's overall state of health.

Studies also show that falling into one or more marginalized group can be even more detrimental. And people who live in stressful environments, take on stressful roles, and meet discrimination in the many places they inhabit receive the greatest hit. Marginalized groups like BIPOC women or ethnic and racial minorities living in underserved neighborhoods face a constant assault from their moment-to-moment environments. One recent study on memory and executive function using data from the international Human Connectome Project, a brain-mapping project that databases and coordinates neuroscience research from all over the world, revealed that racism-based stress, socioeconomic status–based stress, and lower socioeconomic status each have their own negative impact on cognitive function, such that there is an almost exponential effect on those who face all three.[15] Studies like this one consistently prove that the weathering incurred from daily discrimination compounds the effects of socioeconomic status, increasing harmful exposures and creating barriers to the very resources a person needs to thrive.

Epigenetic Weathering

Epigenetic research on social weathering reminds us that our early life experiences can have a lifelong impact on our body's ability to turn on our "beneficial" genes and support our brains to put our minds to good use. One study using longitudinal epigenetic data and brain imaging with teens found that lower socioeconomic status induced epigenetic changes that were linked to increased chances of mental illness later in life.[16] This study found that experiences of low socioeconomic status in adolescence spurred methylation of serotonin regulators and increased amygdala reactivity to perceived threats. Researchers were particularly concerned for teens who had a family history of depression, as they have an even higher likelihood of developing depressive symptoms and disease over the course of their lifetimes.

Looking at racial discrimination, a battery of studies has shown how experiences of racism early in life damage the brain and body's epigenetic response to toxic stress. A review of the literature on racial chronic pain disparities, for example, found that across multiple studies Black Americans had a higher incidence of methylation of stress-response genes and immune-response genes.[17] These studies found that racial discrimination compounded with adverse childhood experiences and low socioeconomic status led subjects to experience debilitating pain and suffering that eventually prevented many from living a normal life.

A global review of epigenetic research and gender disparities found that certain early-life and adolescent gender-based stressors can trigger noxious epigenetic changes in the developing brain.[18] For example, mistreatment and abandonment of female babies and young girls in certain places has led some to end up in destitute environments where they are more likely to suffer higher rates of methylation from their earliest days of life, exponentially impacting their developing brains and their ability to learn. Studies of BIPOC low-income women who have developed depression later in life have similarly shown this to be traced back to harmful epigenetic changes that were incurred in their earliest months of life.[19]

Pillars of the Community

If only we could cure all our societal ills with gene edits and miracle drugs, how simple our lives would be. But no matter how safe and efficacious individual therapies may seem, they don't offer a solution for the disparities that harm people and create barriers to realizing their potential. Our shared environment encompasses all our inborn and innate neurology, shaping our experiences, our exposures, and our epigenomes. The pillars of our economy, our education systems, and our healthcare systems condition the environments we find in our communities. And so, without the option of a quick fix, we must focus our attention on reforming these social structures.

If the global pandemic has taught us one thing, it is that our health and well-being, and even the heartiness of our mental and physical being, is interdependent on that of our fellow humans. Our environments are connected through a vast ecological web. The functionality and quality of our businesses, our schools, our medical facilities, and our homes rely on the functionality and quality of other people's businesses, schools, medical facilities, and homes. Even those who live halfway across the world from us have an effect on our well-being. Our pillars rely on their pillars. If some are at risk, all are at risk. There is no "my sake" and "their sake," just "our sake."

While it might not seem possible to reconstruct every I-beam in every piling of every social structure around the globe, there are ways to retrofit locally and bridge out. The key is affecting *institutional* support for behavioral change in your community. Are healthy, affordable groceries accessible to all? Are there parks and playgrounds where children can safely play and socialize? Does everyone have access to quality education, child and elder care, and health care? Are there programs in place to address the disparities brought on by systemic racism, sexism, classism, and other forms of discrimination?

The best way to accomplish quick and lasting change is to modify policy. That can mean anything from messaging your local legislator or signing petitions to engaging in community organizing or get-out-the-vote campaigns.

But a lot can be done alongside political action. You may be surprised by the extent of local organizations that are already working to bring such changes about. In communities everywhere, environmental justice organizations are convincing local shopkeepers to carry fresh fruits and vegetables and to stop carrying malt liquor. There are nonprofits and not-for-profits lobbying city governments to tighten carbon emissions requirements and install exercise facilities at local parks. There are faith-based organizations and secular ones, too, that are dedicated to bringing growth, mindfulness, and connection to child-rearing and education. There are antiracist and anti-bullying programs that teach new ways of seeing others in schools and work sites. These changes not only raise the standard of living in your community, but also provide models for other communities to modify their policies. They therefore incite a better environment for everyone's genes to develop, express, and flourish in an ever-widening arena.

Community and political action work, but they're not the only game in town. Outreach and education are other ways to put the domino chain reaction into effect, changing hundreds, then millions, of minds. Informative broadcasts of ideas for new relationships with each other and our institutions, disseminated with open-source websites and social media, can offer much-needed inspiration for constructing a new social order. People are spending more time than ever engaging with new ideas thanks to the access provided by digital technology. Your personal vision on a more ethical and nurturing society can quickly become a shared narrative, shared with and valued by people afar.

In education, we must focus on changing curriculums, whether altering canonized literature or foundational definitions and concepts. Social health requires educators to implement a new paradigm. The education route works especially well when the new paradigm has infiltrated key assessments that level you up to fulfill a major, graduate, or get into grad school or a profession. Just think of those gatekeeper tests like the SATs, GREs, LSATs, MCATs. All are places in which the new paradigm can be inserted and leveraged to ensure a measurable increase in awareness.

Though it takes a certain platform to change the specifics of curriculums, you may be surprised by how constant curricular design efforts run, and more importantly, where new curriculum originates. In all my years as a professor, from the Ivy League to an elite medical school to a top-level public research institution, I have consulted with members of the communities in which I worked and lived to source fresh ideas. The genetic literacy projects I have worked on for the U.S. public school system have also brought in community representatives and consulted interested members of the public to figure out how to tailor information to specific populations of youth. And don't forget that every community has elected school boards, and in many cases, people are running on curriculum-based platforms. Communicating your ideas to prospective board members can have a great impact on your local community and can be inspiration for education systems beyond.

In the end, there are as many ways to effect change as there are cares and commitments. As a professor and public speaker, I have focused on using my voice and my writing to raise consciousness. But I have also made it a point to transcend my immediate professional environment to more directly engage with people in my community. In one case, I worked with a museum to create an exhibition on racism in our state. This display fit into a larger exhibition on the history of race and its relation to pigmentation science that community leaders and local schools attended. In another, I worked with evolutionists, genealogists, and museum curators to create an innovative curriculum on gene-environment science that gets K–12 students to think differently about family, personhood, diversity, and humanity. This curriculum has been since implemented at schools all around the country, having effects on teaching and learning, but also family dialogues and self-understanding in communities far and wide.

Even these small actions have planted seeds in a multiplicative way, laying the nurturing ground for the betterment of all of us. If we really want to get smarter, we will need to ensure that there is a way forward.

CHAPTER 9

Seeing Value in Us All

For there is always light if only we're brave enough to see it, if only we're brave enough to be it.

—Amanda Gorman

One of the most devastating things about the score-based paradigm of intelligence is that it ranks us against one another, making some of us seem intellectually valuable and others of us not. The intelligence paradigm I am offering does just the opposite. It asks us to see the learning potential, and therefore the value, in all of us. It marks everyone as learners and as knowers. It also makes us aware of how much we depend on one another—not for our rank-comparison score but rather for our own intellectual development, and especially for our collective, communal growth.

To make way for the new, we must start by ridding ourselves of bad-faith evaluators once and for all, especially the ever-fraught IQ test. Intelligence testing is so ingrained in our culture that it almost feels like a natural, or at least an inevitable, part of our development. But just as there was a world before IQ, there can be a world after it. Seeing the alternatives can help us imagine a better way forward that cultivates the light in us all.

IQ Canceled?

I often wonder how intelligence testing has made it with us this far when it has come under fire for racism, classism, and all other kinds of bias criticisms for decades. It is especially hard to imagine given that IQ testing seemed to be losing its stranglehold over the American education system when I was entering school in the early 1980s. That California court ruling that I discussed in Chapter 2—the one that determined that the California State Board of Education was unlawfully misusing IQ tests in tracking Black students for special education—had already decided that IQ tests could no longer be the sole determinant of intellectual disability for Black children. And that court ruling actually came on the heels of two other class-action lawsuits that questioned the legality of intelligence testing for BIPOC students. The first of these cases, *Hobson v. Hansen,* came to trial in 1967 when civil rights attorney Julius Hobson sued the Washington, DC, Board of Education (and school superintendent, Carl Hansen) for using test scores to track low-income students and students of color for special education (programs that they found tracked students not only for subpar education programs but also for low-paying labor later in life).[1] The judge in this case ruled that the tracking under question was indeed biased, and that the DC Board of Education would need to begin monitoring the racial breakdown of its tracking in order to make sure it was not disproportionately assigning Black students to particular educational tracks. The second of these cases, *Diana v. State Board of Education,* came to trial in 1970 when civil rights attorneys Marty Glick and Maurice "Mo" Jourdane sued the California State Board of Education for using English-only intelligence tests to track students for whom English was a second language.[2] The judge in this case ruled that intelligence tests had to thereafter be administered in a student's native language.

Sadly, none of these rulings undermined intelligence testing in the American public school system. In all of these cases, the judges found fault with the application of test scores in tracking students for special education. They did not find fault with the tests themselves. Even in the California case that eventually ruled that tests could not be the sole determinant of intellectual disability for Black students, tests were still an option if used in conjunction with other data with permission of the court. In the years following the 1986 ruling, test makers attempted to rid IQ test questions of cultural biases, and they successfully convinced parents and educators to bring the matter to trial again. In 1992, the California judge responsible for the injunction on tests rescinded it.[3] Intelligence testing was reinstated in California as the appropriate measure for educational tracking and has remained so ever since.

Proving Disability

At the same time as all of these court cases were playing out and in the years leading up to the present, the U.S. government has doubled down on its requirement that schools and parents use IQ scores to prove disability status. Since the Individuals with Disabilities Education Act (IDEA) went into effect in 1975, educators have been required to evaluate children with standardized assessments before placing students into special education classes.[4] An IQ score that falls at least two standard deviations below the mean (75 or under) also enables a student to receive intellectual disability status, which mandates that the student receive accommodations and modifications in all spaces of education. This policy, which has been fortified with each new Disabilities Act, for the most part accounts for the revival of intelligence testing across American education.

IQ tests therefore function as a gatekeeper to rights and resources for anyone who is not thriving in the educational mainstream regardless of whether their problem is decidedly cognitive or not. And this makes for a catch-22 for kids who have already tested low and been tracked for special education and disability services. IDEA only covers youth until they reach eighteen. Once a student turns eighteen, they must maintain intellectual disability status in order to continue to receive accommodations and modifications in their new learning arenas, as well as whatever services they benefited from in their old, under the Americans with Disabilities Act. At this point, if they want to go to college, where they most likely will need accommodations and modifications, and where services will be of great assistance, it is actually in a student's best interest to retain their former IQ score so that it may confer disability status as they move into higher education.

With these policies in place, test makers say that we need our IQ measurement system to serve youth who would otherwise have no access to disability rights and resources. And, as the laws stand, they are right. But I think of my own experience of testing lower in vocabulary while testing higher in other areas. As a child who had a solid handle on shapes and numbers but—as a member of a bilingual household—lacked exposure to certain words, the tests I was given couldn't provide an accurate assessment of what they were meant to gauge: my ability to reason. So how can we trust them to do so for students who do not present as neurotypical, or who are facing various other challenges in their lives, especially in a country plagued by social and environmental inequality?

The story of one of my college students who also came from a bilingual home, yet one in which English was even less prevalent, illustrates this conundrum even better. As a child, he scored so low on an IQ test that he was labeled "intellectually disabled." He was separated from his peers in the mainstream learning environment of his school and put into special education classrooms. After a decade of listening to his teachers tell him that he and his special ed classmates would never go to college, he almost gave up hope. But a member of his community suggested he try a course at his local community college, and in that new educational environment he thrived. A few years later he was in my classroom at Rutgers University, acing his exams and performing at the top of his class. He was charting a path for the career of his dreams and several graduate programs were already interested in him. His IQ score was just plain wrong. And it almost cost him his future.

The statistics support my student's experience. And they are positively harrowing. When the class-action court cases that cried "racism" at tracking children by IQ were being waged in my youth, Black and Latinx children made up less than a quarter of the student body but accounted for the vast majority of the student population labeled "mentally retarded." Today, Black and Latinx children are still more likely than any other racial group of children to be labeled as "intellectually disabled," and both groups of children are disproportionately tracked for special education in their schools.[5] Black youth are also disproportionately labeled as having an "emotional disturbance" (such as what we educators called "severely emotionally disturbed" when I was teaching K–12 to put myself through graduate school).[6] They are more likely to be harshly disciplined by educators in their schools and passed on to the criminal justice system when a mental illness is diagnosed than any other group of children. Meanwhile, Asian and White students who have been identified as intellectually disabled are the most likely to stay in school and receive a regular high school diploma.[7] White students in particular are disproportionately "mainstreamed" in regular classrooms, spending an average of 80 percent or more of the day in their school's mainstream learning environments than their peers of color.[8]

Why are we settling for such a bogus metric? Do we really want to continue to rank someone's intellect based on their cultural familiarity with certain test questions? Do we really want yet another way to distinguish the privileged as superior, especially when they already possess environmental advantages that propel them to power? I recommend that we abort this mission. Get out while we can. We must say "no" to any privilege quotient that's masquerading as an intelligence quotient.

Direct to Cognitive

Some intelligence scientists might respond that we should keep our admittedly biased tests because they are the best thing we've got. We can use them to establish a baseline for lead toxicity. We can use them to prove that a person who has been classified as developmentally disabled is in fact educable. To this, I say let's use direct cognitive metrics that can give us information about the function of our brains.

If we are worried about memory dysfunction, let's test it directly. Memory tests take just a few minutes and are composed of simple memorization tasks that require no prior familiarity with specific words or numbers. They usually start with a test administrator asking you to memorize a list of unrelated objects or unrelated shapes, say a shoe and a cup. The administrator might also ask you to pick those objects or shapes out from a picture. You might be asked to count forward and backward in your own language and numerical system, or to draw a picture or copy shapes that the administrator shares with you. Again, you will need no prior knowledge of any of the material on the test. In fact, the items on it won't depend upon you having context for them because they will have been selected to be as unrelatable as possible.

There are also memory tests that assess memory processing, for both short-term memory and long-term memory. As part of a short-term memory test, you may be asked to talk about something that happened to you in the last couple of days. For a long-term memory test, you may be asked to talk about something that happened many years ago. The test administrator won't be looking for finesse or poetry in your description; they won't even consider the logic of the storytelling. Instead they will be homing in on your cognitive processing speed and your mental agility in toggling between past and present.

If we are worried about visuo-spatial dysfunction, that too can be assessed directly. We have so many tests at our disposal that go straight to the point, such as spatial navigation tests that prompt someone to maneuver through a landscape or topography. Even mazes like the kinds that scientists use in animal models can tell us a lot about brain functionality in particular zones of cognition. And these tests provide concrete numbers on functions at work inside our brains that can further illuminate problems that need immediate or long-term medical attention.

These tests are so much more accurate at diagnosing impairments than IQ tests that it seems strange indeed that educators aren't more incentivized to use them as their primary means of analysis. The reason, I fear, is the financial bottom line. Even if IQ scores weren't the official mandated marker of disability status, IQ tests would be preferred because they are cheap and convenient. And they are cheap and convenient because they are standardized. They do not require that an expert administer the test, and they most certainly do not require one-on-one interaction. They are all-around cost-effective. They cost a school system less time, less thought, and less care.

Just think of the difference in terms of labor, time, and space. Aptitude tests can be given in the same way that any standardized test is given—anywhere, by anyone, on any computerized device. When I was a kid, our vice principal or our counselor would open the test, the librarian or another staff member would oversee it, and the teaching assistants would handle the rest (shuttling me to the restroom or helping me sharpen my pencil—these were the days of paper booklets and printouts). Once the timer started, all of us test takers sat in our seats, eyes downcast, privately toiling away at the handouts on our desks. We were explicitly barred from asking questions or making a peep.

The school didn't need to have a psychologist on hand to work with kids like my schoolmate Charlie, who was eventually given disability status, in a room alone apart from the rest of us. The school didn't need to test us kids one at a time, in sequence, in a patchwork of different seasons throughout the school year. And when some of us got antsy while others stayed focused? The school didn't need to question whether there was something else going on for the uneasy. The school didn't need to stop the test and deal with that student directly and personally. To the contrary, a fidgety student only confirmed the guiding prejudice that infused not only my school district but also the entire American education system that only some of us kids were intellectually bright. Only some of us were capable and a very limited few were "gifted and talented," valuable beyond the norm, rising stars on our way to light—and hold power and rank within—the world.

Putting Out the Lights

My school was on a tight schedule of aptitude testing. Testing an individual kid for cognitive function one-on-one in a personalized way throughout the school year wouldn't have been feasible given the LA Unified School District's budget crisis and the stressful atmosphere that underfunding created. Our administrators and teachers were volleying between rallies and strikes, so they had a lot of material to squeeze into an increasingly short amount of time. Because my mom worked way before and way after the school day took place, I still attended all my classes. I did the dittos and workbooks that the substitutes gave us to keep us on track. But with other students so frequently out getting temporary homeschooling from their stay-at-home parents, my classrooms were always starting over, getting back to square one to prepare us for that test that was on our horizon.

In my school, the teachers were also being evaluated by our scores. Time and again, the school administrators would say, "Don't worry, this test is just helping us make sure that the teachers are doing their job." This did not dispel our anxiety. If anything, it increased the pressure on us—worrying for our teachers that our aptitudes were a reflection of theirs.

I wish I could say that my experience was unique, but I've seen a similar phenomenon throughout my years as a K–12 educator, and now again as a parent. When I graduated from college in the early 2000s, I entered into an AmeriCorps civil service program in the Oakland Unified School District. There I worked in schools that had been identified by the district as its ten lowest-scoring schools. For nine months, I worked in one particular elementary school with their K–4 students, pulling kids from class who had not yet learned to read and working with them one-on-one to bring them up to speed. Meanwhile, the teachers diligently did their part to prepare the students for the standardized test ahead.

The learning environment was severely stressful, and the

children were even more anxious than the educators, who were also being evaluated by those scores. All but a handful of the students were Black children who lived in a neighborhood that had been ravaged by predatory policing and police brutality. Many of them were descendants of former Black Panthers who had been unjustly incarcerated, and a good portion of them did not live with their parents or their family in a stable home, let alone a home that enabled them to do homework. Nearly all of the students were getting three meals at school from its crumbling cafeteria (menu items like sugary cereals and microwaved hamburgers, nutrition-poor foods that hardly promote health let alone give a child the raw materials they need for learning). The ceiling panels were literally falling on our heads. But it was the pressure of the test that so palpably stressed them, and the value system that was set against them that so clearly hurt them. They saw their worth—their light—wrapped up in their test-taking ability, and it was an insurmountable task.

I was relieved to find that most of the teachers at the school were critical of the testing mandates, and many regularly expressed frustration to the principal, the staff, and to me. But their school, like every other in the country, was under pressure to prove its worth using its students' test scores. The school was under threat of being defunded, so it was all hands on deck to boost those scores in order to prove to the state that there was enough light in their midst.

My job was supposed to be simple: help kids read. But it was in fact much more complicated. I was under the same pressure to get them up to a specific level marked by a specific score. When it took longer with some kids, I was asked to confirm that they were "special needs" or "educable mentally retarded." It served the school better to be able to exclude those children out of the mainstream testing population.

I remember working with one student in particular, whom I'll call Monique, who when quizzed on the name of any letter of the alphabet replied "G." I changed tactics and began working with her on seeing the phonetics in the letters and then linking those sounds to Monique's favorite things—ants, bees, cats.

I nearly jumped out of my seat the day that Monique read her first word: *sun*. I wanted to hug her when she then read a series of rhyming words: *b-un, f-un, r-un*. She was proving to me and to all of us that she was motivated and capable and ready to string together all of the individual sounds. She was reading!

It turned out her achievement was not enough for her teacher or the system that constrained them. At the school's spring open house, Monique's teacher took her mother aside and implored her to have Monique identified as disabled and moved to special education. I couldn't believe what I was hearing. Before I could think about it, I jumped in and stood up for Monique. I blatantly disagreed with her teacher and told her mother that Monique was just as capable of reading as any of us. I said, "It just takes more time and care but she's doing it!"

Her teacher, who was practically fuming from her ears, rapidly came back with what I thought was the harshest and most unfair of words. In front of Monique and me, she said flat out that Monique was "uneducable." But what killed me was that her mother agreed. She simply shrugged and in a low, tired voice said, "Makes sense. My mother and her grandfather and her father were like that. None of them could do school right."

I learned from that experience how disposable children can be to a system that is predicated on standardization and scoring. I also learned just how deliberately this system penalizes the people who need help the most. This can dampen a student, her teacher, a whole school, and a neighborhood. It can pull the plug on someone's dreams and potential, depriving them of bright livelihoods and life chances.

Pay to Play

I've thought a lot about Monique lately, as my twins are entering the public school system after having been sheltered in a play-based preschool for the past year. Here in the state of New Jersey, aptitude tests are administered to certain K–12 grades at the start and finish of the year, and given to all students in grades three through eight by the end of the year. Though I'm certain their teachers do not want to spend the year "teaching to the test," it is inevitable that they must prepare my kids and all the students in the school for the material on the test. That means using everyone's time to focus on test material and not other kinds of material. It also means that my kids and their peers are losing out on the opportunity to learn in the progressive, collaborative way that is best for their developing brains. They are losing that child-led, student-centered, inquiry-based form of learning that has been the basis of their lives up until now.

A recent study examining the effectivity of free, public pre-K has shed some light on the detriments of the score-based paradigm and standardized testing, especially in terms of the loss of student-centered learning that they provoke.[9] In a decade-long longitudinal study of Tennessee's statewide pre-K program that was designed to help students from low-income backgrounds, researchers found that this program, however well-intentioned, had a significant negative effect on them. Instruction was focused on basic reading, science, and math, using a teacher-centered learning style in which students were lectured for long periods. In the end, researchers found that this model resulted in students progressively testing worse as they moved through their K–12 education.

When examining what went wrong, study scientists found that there were two major problems with the standardized program. First was the learning style. The children in the public pre-K program were not just being taught to the test but also to the testing. In other words, learning revolved around getting students "school ready" by having them be "testing ready"—having them learn to be quiet, sit still, receive information passively, and then study and perform aptitude of one's basic skills the way that they would have to in their K–12 test-driven school environment. While the program's students finished the pre-K year scoring higher than nonprogram students in school readiness, their following aptitude tests taken in each subsequent grade in elementary school saw progressive declines in all subject areas. In addition, program students were more likely to suffer disciplinary action at the hands of teachers and administrators, as well as get suspended from school. The scientists determined that what these program kids were missing was the very play-based education that higher-income families were privileged enough to give their children. Instead of talking at the students and preparing them to endure testing and score high on tests, they should have been letting the kids do the talking, the asking, the investigating. They should have been providing them an education rich in movement, nature, art, and play.

The second problem was a bigger structural issue. The public program was housed in the very K–12 school buildings that students were set to feed into. These buildings were also teacher-centered, with classrooms set up to facilitate lectures and testing, and with bathrooms, eating facilities, schoolyards, and lockers a far walk down the halls. The study scientists found that the program kids spent most of their day "transitioning," that is, moving through the buildings. Unlike private preschools, they did not have rooms designed for small children and their needs, such as having outdoor play yards, eating spaces, cubbies, and bathrooms connected to the main indoor learning space.

This lack of structural support reduced the amount of time that students actually spent learning, but it also had a negative effect on their behavior. Teachers were constantly reminding students to line up, be quiet, not touch the other kids around them, walk silently, look ahead only, and on and on. So, the transitioning time that dominated the children's daily experience (and that ate up all their learning time) was in fact discipline time. The scientists on the study concluded that the children were learning that school was a place of external control, a place not to be trusted. At the same time, they were not learning internal control. They were being told that they themselves couldn't be trusted, and that put them in an unworkable position.

These researchers' takeaway was that introducing the K–12 score-based, standardized culture of education was harmful to these children. Based on their prior decades of research, they additionally surmised that small children are developmentally "allergic" to the learning style and disciplinary habits that are embedded in it.[10] Their conclusion was that early childhood education should be a place of curiosity and discovery, a place that stimulates a child to learn about themself in connection to their environment, and to thereby grow intellectually and emotionally. It should not be a place where a child learns that their only value is in performing their aptitude.

This research begs the question: When is the right time to kick-start standardization and scoring? The scientists here look to other free, public pre-K programs that have done better, and they have found that some of those better-funded programs have avoided the bereft teacher-centered learning style and standardization that the Tennessee program succumbed to. As a result, they recommend that school districts going forward hire teachers that are well versed in play-based, student-centered early childhood instruction, and not teachers who are just coming off a lengthy march toward K–12 certification. They suggest we postpone that score-based, standardized education for another time.

Still, looking at my own kids' path ahead, I fear that this question poses a fallacy. Instead of asking when, I want to know: Why? Why would I, as a parent, or my kids, as emerging learners, want this kind of curriculum? Why would my kids need to be testing and test-prepping instead of creatively and collaboratively learning?

Another way of asking the question: How? How does this benefit my children? How does this grow and develop them to their best potential? While I can imagine a number of reasons to assess my children's portfolio of recent work in order to measure their own progress against their past performance, and while I would want my children to be tested for specific learning challenges such as dyslexia or "information overload" if they appeared to be struggling, I cannot imagine how it would benefit them to measure their so-called "aptitude" against another child's aptitude, nor all children of their classroom, their school facility, their district, or their age. I would much rather have them working their growth mindset, mindfulness, and connected-learning muscles, learning and building their knowledge base by asking the questions that strike them as important, as they progress through elementary school.

The pre-K study's findings also raise for me another set of issues that we have learned about from a different longitudinal study's findings on raising kids with curiosity and inquisitiveness. In research that stretches back to when I was in elementary school but has been tried and tested in the decades since, sociologist Annette Lareau found that disciplining kids to sit quietly, behave, accept what they are taught, and do what they are told only hurts children. Instead, teaching children to question the world around them, including authoritative knowledge and what figures of authority say, helps children see that they are active learners, they are leaders of their own destiny, and their curiosity is what counts.

Lareau, who focuses on the parenting end of a child's education, calls it "concerted cultivation," where parents supplement school learning and teach their kids to be critical thinkers who think and speak for themselves.[11] In the context of school learning, and in the interests of moving us away from the score-based paradigm that dominates it, I want us to focus on the self-driving force of it. Supporting kids to be curious, creative, and inquisitive strengthens their self-reliance and their self-driven nature, and it enhances their sense of self-worth. Supporting their critical thinking and executive judgment also helps them to tackle the decision-making responsibilities that they face across the spectrum of what they are engaged in in school and outside of it.

As Lareau reminds us, this isn't something that expires when our kids turn five or seven or even seventeen. Teaching our kids to use their own wits and to draw on their wider bases of knowledge that they have acquired outside the classroom in order to think creatively inside it is something that we should be doing not just in preschool but for the K–12 years beyond. It helps them to grow into self-driven, internally motivated adults who can chart a unique course based on their own goals and interests.

Closing the GATE

Coming back to the matter of IQ testing specifically, there is another argument that test makers and IQ proponents make for having schools score children by IQ. Many argue that IQ tests help on the other end of the spectrum with identifying and tracking the "gifted and talented" for a better education than they would otherwise have. They suggest that eliminating IQ tests would squash an underprivileged kid's chance to shine. Contrary to seeing IQ tests as detrimental, this camp champions IQ tests as a most critical weapon in the fight for social justice, because without establishing a high score, bright children can just fall through the cracks.

Under the constraints of our current system, they are completely right. Just like we saw in the case of proving disability status, if college prep like "gifted and talented education" programs, or what many American school districts call "GATE," is tied to their aptitude score, then students need that score to access that higher-quality education. Indeed, it's true that in American public schooling, the GATE to college only opens to those who have established a higher aptitude than the mean.

But we must ask ourselves: Is this the right system for educating our children? Is it right that the education system doles out higher-quality education based on aptitude scores? Or should the GATE to college be opened to those with lower scores or average scores if they perform well in different ways? And what about the bigger question lurking behind all these: Should score-based programs like GATE even exist in the first place?

The New York City school system, which some of my family members work in and which educates many of my children's closest cousins and friends, has been struggling with these questions for some time. In 2021, then-mayor Bill de Blasio announced that he would close New York City's GATE program because it was producing a racially biased education system.[12] In the preceding years, Black and Latinx students, who made up the vast majority of New York City's student body, were getting only a minuscule percentage of GATE seats. Meanwhile, White and Asian students were being tracked for a higher-quality curriculum that was leading to college, post-college education, and white-collar jobs. Far from helping identify rising stars, GATE was further stratifying an already stratified system.

Kindergarten was a particularly problematic example, given that GATE has been proven to help propel kids for even higher amounts of educational and job success the earlier that they enter the system—in other words, the sooner you open the GATE to college, the more effect the track has. From 2018 to 2019, the White and Asian kindergartners, who made up less than a third of the kindergarten student population, held more than two-thirds of New York City's GATE kindergarten placements. De Blasio wanted to end GATE and instead offer all students the choice of testing into an accelerated learning program at the end of third grade.

De Blasio was not able to see his dream out during his tenure as mayor, and as a result New York City's GATE debate is ongoing. As of this writing, Mayor Eric Adams had not closed the program, and the future of gifted programming in the city schools remains unclear. Still, there is a lot we can learn from this example, starting with the economics of programs like GATE.

New York City's gifted and talented program had been expanded in the early 2000s by then-mayor Michael Bloomberg as a way to entice affluent White families to stay in their zoned school districts rather than send their children to one of the city's many elite private institutions. It was not expanded evenly across the boroughs, and it was left to wither in schools that served lower-income families. Bloomberg also created New York's own standardized aptitude test to both narrow and simplify admission to GATE. As a result, privileged families could more clearly see how to prepare their children for the test, and many hired specialized tutors for this purpose. This had two effects. First, the margin between White and Asian students and Black and Latinx students grew. Second, the district could boast of an uptick in the numbers of kids going to college and excelling in the profession of their choice.

Just like in my school and in some of the San Francisco Bay Area schools in which I later taught, the New York system was "fixed" to reward high scores that could garner more funds for individual schools and the entire school district. And it worked. In my case, it was a great asset for my district to have a high-scoring kid like me remain in the public school system as opposed to going on scholarship to my local private school. It was even more important to my elementary school that I didn't go to the local public magnet school and take those funds with me to them. My school and the district literally needed my scores to be high, just like they needed Charlie's scores to be low enough to provide him the right scoring accommodations to ensure he wouldn't mar the school's mainstream testing stats.

My elementary school turned out to be a success story in this score-based financial reward system. Throughout the 2000s, it enriched its GATE program, and was later converted to a charter school that has a thriving GATE community. Meanwhile, in the Oakland Unified School District, where I taught, the elementary schools have a different story. With fewer than 20 percent of students scoring at or above state standards, nearly all of them have been shuttered.

Reevaluating the Evaluators

As is clear from the state of schooling, and testing, in the U.S., we are locked into a devastatingly unequal educational infrastructure. We have a system that rewards scores and not effort. It lets aptitude scores determine whether a school invests its limited quality education resources in a student. It determines a student's worth based on an incomplete snapshot of their ability, and not their actual need or potential.

To me, this system is completely upside down. We are all valuable. We are all worth investing in. We just have different needs.

Rather than evaluating our children with a blunt-force test, I want us to identify individual needs and meet them accordingly. I want the children who need language instruction to get extra help with their vocabulary and reading. I want the children who never had a math tutor to get tailored coaching in math. I want the kids who never had the privilege to do a science experiment at home to get first-rate science education. I don't want their quality education to come at the cost of their scores.

I am particularly passionate about this topic when I consider my own kids and their similarities and differences. Intelligence scientists who promote the false narrative that IQ is genetic would posit that my twins have identical aptitudes. Many also assume they have identical strengths and weaknesses, and therefore identical needs. They say that an IQ test of one is the same as an IQ test of the other, and that their school need only tailor their education to it. But I hope that we can see the fault in these assumptions and recognize the uniqueness of my kids and all of our children. I dream that we can move away from scoring their intelligence as if it were preprogrammed and preset, and make strides toward giving them their rightful place at the center of learning. Only then will we know what they need most, how we can empower them, and how we can ensure that they are able to light the way forward.

Conclusion

When I was a kid, I believed that my academic and test performance summed up my ability. I was no more than a score. When that score clocked in high, I trusted that it could be my ticket out of an environment that did not nurture or provide enough for me. And in many ways, it was.

Today my understanding of and relationship to intelligence is far more complex and complicated. I know that it is private yet public, personal yet political, inside yet outside. It is something inherent, innate, essential—yet malleable, improvable, plastic. It is relational and interdependent.

It is also a cultural narrative that needs to be overhauled. Our narrow definition of intelligence is the legacy of racism and patriarchy—of men who created and propagated the belief that only some of us were genetically endowed. As a result, we have been playing and replaying the same mistakes—and all of us are suffering. It is time for a new outlook and a new framework for valuing and caring for ourselves and each other.

Reimagining Intelligence

In the intelligence paradigm I am calling for, we are going to need a completely new definition of, if not language for, intelligence—a way of talking about intelligence that connotes its universal, democratic nature. We *all* use our minds to interact with our environment. We *all* grow, learning and developing our brainpower. We *all* have the potential to see and capitalize on our mental fluidity, to become more mindful, and to learn collaboratively with others around us in ways that can improve the collective good. This is true no matter who we are or how "neurotypical" we may seem, no matter where we have started from and how privileged we have been in our early life, and no matter where we find ourselves presently.

Sure, scientists have already attempted to complicate the word. Many adhere to some variation of Howard Gardner's multiple intelligences framework, which assumes that there are many forms of intelligence beyond the kind measured with standard IQ tests.[1] Some also cite Daniel Goleman's emotional intelligence model in maintaining a distinction between rational and emotional intelligences and a belief in the power of our emotions to condition our ability to reason.[2] Even many modern-day proponents of IQ testing, who believe that a person's aptitude can be captured in a single score, have come to believe that intelligence is best split into at least two basic forms: "crystallized" intelligence (a store of knowledge that comes from our past experiences and exposures) and "fluid" intelligence (our personal capacity for active reasoning).[3] These scientists work with test makers to design test questions that can capture both kinds of intelligence, which they perceive as equally valuable to a person's intellectual ability.

To me, these attempts to multiply intelligence come up short. Now, I imagine it would be too much of a challenge to eradicate the word, or the notion of, intelligence from our values and ambitions, let alone from our hearts and minds. Or to extricate it from the systems that rely on it to measure our worthiness, to score us against one another. But it is worth thinking about what might be better in its place.

Just as I recommend understanding our mind's genetics in terms of cognition that is attuned to your environment, I recommend understanding intelligence in terms of awareness of that environment. Being alert to and contemplative of your environment is *the most natural and important part of having a human brain and being alive.* To rethink intelligence as awareness can help open up the world to you in terms of growth, mindfulness, connectivity, and collective creativity.

At the beginning of this book, I shared some dictionary definitions of intelligence. All had the feature of defining intelligence in terms of aptitude. And aptitude screams quantity. It screams inflexibility, immobility, rigidity. A fixed essence of who you are.

If you look up the word *awareness,* it by contrast alludes to qualities like insight, perception, sensitivity, and attention. These things are fluid, in process, and aligned with active cognition. They are what we use and what we do all day long, again no matter who we are, where we are starting from, or what our current situation happens to be.

Awareness actually prevents us from thinking about amounts and scores. Rather, it inspires us to realize our inherent neuroplasticity—our continual growth and development. It helps us stay focused on cognition, cognitive health, and powering our brains to mobilize. It helps us move, seize, and grow, and to do so collaboratively in connection with those around us.

Cognitive scientists have many ways of thinking about thinking, and they all emphasize action. One commonly known list of ways we think is called Bloom's Taxonomy. It includes remembering, understanding, applying, analyzing, evaluating, and creating. These are all qualities, not quantities, and in fact they defy rote quantification.

A more recent and highly popularized approach is to understand thinking as either intuitive or deliberative. In *Thinking, Fast and Slow,* Nobel Prize–winning economist Daniel Kahneman explains that we are always toggling between using our brains in automatic, unconscious processing and effortful, logical processing.[4] These too are different kinds of thinking that are equally essential to our action in the world. They are not aptitudes that some of us are born better suited for than others.

So really what we need to do is think about the active processes our brains are engaged in and the quality of those processes. We need to shift toward intelligence as awareness born of our growth, mindfulness, and connectivity, and retool how we consider basic thought processes. Most importantly, we need to examine our environments and ask ourselves how we can improve them. We must start seeing ourselves as intricately connected to the well-being of that environment and, above all, the people within it.

Creating a New Relationship with Our Intelligence

Reimagining intelligence in terms of active, ongoing, limitless awareness has the potential to subvert many social arenas of our current system. Right now, intelligence scoring has us locked into a system of rewarding some and disadvantaging others. We mark a select few as gifted and others as ineducable, and we score people against one another, marking some as better and others as worse. Just imagine how different our lives could be without that hierarchy—especially if early childhood education and secondary education were to discontinue the practice of ranking. If we replaced scoring with growth mindset, mindfulness, and connected learning in a child's earliest educational settings, we could offer that child not only the chance of an enriched education, but also an enriched sense of self-worth.

A new paradigm for intelligence can also empower the many of us who are so very far beyond those early years. Imagine having the ability to create a new relationship to your intelligence—one not based in competition or scarcity but rather one that stimulates growth and self-confidence. We can start by realizing that we are not inherently flawed, no matter how real our internalized fears are, and no matter how much the trauma of earlier events in life reverberates through our current life experiences. We are not the sum of our test scores.

I say this as a person who ranked high under the score-based paradigm and therefore should have been sheltered from fears of inferiority. Even as I watched others around me score lower, I internalized my own sense of intellectual inferiority based on my inability to achieve perfection. I felt inadequate to my core when, as a young child, I scored below the top percentile. I again felt useless when, as a teen, I achieved less than a perfect score on my SATs. There was always someone out there who scored higher than me, even if they weren't in my class. So, while I knew I was protected from the academic shaming that many of my classmates suffered, I feared that I too was innately undeserving in my own way. And I believed that my imperfection belied my relative worth.

Moving into high-achieving school settings and competitive professional arenas only stoked my fears. It seemed I was always so close to getting the honors and accolades that I desired. When I did achieve high, it was never enough. And though I only wanted my colleagues to succeed at their jobs, I was relieved every time it wasn't me that was warned about their progress at work. I strove diligently to add line after line to my CV because I knew how devastating it would feel to be in their position.

That's the problem with a score-based system of rank-comparison: at its heart is an unshakable discord, a brutal conflict between the ranked and within them. The system creates winners and losers, and it pits everyone against each other. This inspires fear and antagonism, at times subliminal and at times explicit.

A score-based system of rank-comparison is even more detrimental when, as in the case of intelligence, it proliferates labels that are supposed to stand in for a person's inner value. Nine-year-old Rosa Marcellino knew this when she and her family fought to change the label "mentally retarded" to "intellectual disabled" in the early 2000s. Rosa, who was born with Down syndrome, felt diminished by the word *retarded,* so her brother and sisters and her parents petitioned to have the label eradicated. Their work was part of a larger movement to "End the Word" so that one day neuro-atypical people would be seen for their unique intelligence and their equal value.

Unfortunately, Rosa and her family's dream has far to go, because our beliefs about what is "normal" and "extraordinary" have so insufficiently changed. Not enough has been done to challenge labels on the other end of the spectrum, like *gifted,* and so the false dichotomy of intellectual winners and losers persists, linked to a deeper belief in our inner merit.

When you think about it, our present-day intelligence scale is little better than Plato's pyramid, which he conjured up using nothing but his imagination (and his social privilege) thousands of years ago. Plato didn't "do the math," because "the math" didn't yet exist. What did exist, and what sadly continues to endure, is an intelligence scale born of pseudoscience that is used as proof of innate inequality. We continue to back a social order built to substitute our intelligence labels for our inner worth, and we continue to allow particular labels to be attached to particular items of external worth—such as quality education and job placement and hiring. So, on the one hand, we cheat the majority of people who have scored lower on the curve (those ranking average plus those ranking lower than average) out of the highest-quality resources. On the other hand, in giving the minority who have scored higher on the curve a better deal, we ensure that the system keeps rewarding that minority and cranking out better outcomes for them alone.

What's worse, a score-based system of rank-comparison marks a space as a territory of competition when that space is supposed to be a nurturing ground for its inhabitants to learn and grow. This means that for anyone who buys into the system, whether by believing in their relative superiority or by believing in their relative inferiority, they all experience an unhealthy state of anxiety that distresses and distracts. As my own experience evinces, the so-called "gifted and talented" end up constantly comparing themselves to each other and stewing in the fear and shame that they are still relatively worse off than someone else. Meanwhile, the marginalized are continually subjected to an onslaught of negative judgment, harsh treatment, and withholding of learning or promotional opportunities, which can become a self-fulfilling prophecy that holds people back from their true potential.

Creative collaboration, let alone clear thinking, cannot thrive in this kind of environment.

All of us have come up under the score-based paradigm, so very few of us will have escaped the trauma that it has engendered. But we can begin healing, starting with ourselves and our own relationship to intelligence. We can reframe our "wins" and our "losses" as a false dichotomy, we can reject rank-comparison where we encounter it, and we can begin valuing ourselves (and *all* others) as the innately intelligent beings that we are.

From Hive Mind to Thrive Mind

No matter where you came from, where you grew up, where you settled (even temporarily) to live your day-to-day, if you are reading this book, you are already thinking smarter. You are thinking outside the box. You have shown interest in shifting your ideas, priorities, and behaviors to create empowerment for yourself and others.

You have come to understand what makes a human brain and body tick, what powers your epigenome to power your genome to power your neurons to power your mind to think. In carefully and critically considering the various influences of your personal life and relationships, your places of learning and work, and the social structures of the world near and far, you have also demonstrated a keener awareness of your environment.

Many if not most people around you will still assume

that intelligence is a concrete quality you are born with. That there is a genetic lottery—you're either lucky or you're not.

By divesting in that stultifying story, you are moving the needle in a new direction. You are setting yourself up for success, and in the process helping others to maximize their environments. You are moving from hive to thrive.

We are not all the same. We are born into different families and caregiving frameworks; we are born with vastly different amounts and kinds of resources. In our minds too we are different (even those of us who share a genome).

By virtue of these differences, we need each other. We need to become more sensitive to what's going on around us, more able to see things clearly, and more capable of using our insights to put our attention to and process new information. Our responsibility to each other is our responsibility to ourselves. We need to see the infinite potential in all our minds.

Acknowledgments

This book grew out of the work of many caring researchers, writers, and visionaries of a number of communities that I am fortunate to belong to, and I am grateful to all. My deepest gratitude goes out to several of my closest soul sisters—Ruha Benjamin, Erika Hayasaki, Janet Vertesi, and Alondra Nelson—who helped me grow my ideas from seed to tree. Their love and generosity as I have staked out a new public dimension to my work has been invaluable. I have learned so much from this phenomenal intellectual community, and I am humbled to be a part of it.

The genomics community, including the geneticists and epigeneticists who have opened up their labs to me, helped me to see the intricacies of human genetics and its relationship to the broader environment. I especially want to thank Vence Bonham and Francis Collins at the National Institutes of Health for paving a way for genomics to become a full-fledged gene-environment science that grasps the power of social environments to shape our minds and bodies. Their commitment to prioritizing the public health impact of genomic research has set us on course for building a healthier and more equitable world.

Thanks are also due to the CRISPR community, the Precision Medicine community, and my fellow researchers working with the National Academies of Science, Engineering, and Medicine. Françoise Baylis, Aaron Panofsky, Aravinda Chakravarti, George Church, George Daly, Jennifer Doudna, John Evans, Gil Eyal, Jeremy Gunawardena, Ben Hurlbut, Sandra Soo-Jin Lee, Pilar Ossorio, Janet Shim, Charis Thompson, and many others have worked to ensure that social and equity debates have featured front and center in CRISPR and Precision Medicine forums from Day One.

This book started from a conversation with the brilliant ideas editor and literary agent Georgia Frances King. Georgia took my many passions and helped me draw them into a forceful literary form. She is truly unique in this world, and she puts her whole heart into everything she does. For that, I am grateful.

This book has been crafted with the artful insight of Julie Will and her amazing team at Harper Wave. Julie has gone above and beyond the role of an editor; she has been my key interlocutor. From the very beginning, she has pushed me to make my work stronger, clearer, and more relevant. My deepest thanks go to Julie, Amanda Pritzker, Yelena Nesbit, and to the HarperCollins powerhouse for bringing truly mind-opening literature and science that can touch lives to our world.

My literary agent, Will Lippincott, has deeply influenced me. It's impossible to translate the generosity of his spirit into words. He has held my hand, steered me, and shown me kindness at just the right moments. I am so thankful that he is at the heart of my literary family.

I want to also thank the fearless boss babes at BMD and Outspoken. Kristin Steele, Catie Bradley Shea, Tara Berthier, and Tori Marra have all helped me find my voice, share my message, and reach people outside the halls of science. This book was an iterative development that grew out of my conversations and work with them.

A big shout-out to the members of my writing group. Ruha Benjamin, Bettina Judd, Keisha-Khan Perry, Ashanté Reese, and Bianca Williams have all kept me on task and inspired me to sing loud and stretch wide from the earliest inklings of this project.

Another shout-out to my Princeton writing group: Ruha Benjamin, Emily Merchant, Janet Vertesi, and Sonia Van Wichelen. Viruses, campus security, the freezing cold . . . nothing has stopped us from showing up for each other when we've needed it most. Beyond giving it up for me in intellectual terms, they have helped me balance life and work, and they have shown me how to be a better parent. I am so lucky to be growing and learning with you!

I want to thank the writers I write with—Torsten Heinemann, Catherine Lee, Ann Morning, and Wendy Roth—for supporting me in research, art, and friendship. No matter where we are in the world, I am fortunate to always count on you.

Thanks go to Rutgers University and the Rutgers School of Arts and Sciences, and especially to my colleagues in Sociology. With the support and collegiality of fellow faculty members like Catherine Lee, Norah MacKendrick, and Joanna Kempner, I have felt warmly welcomed into the Rutgers fold.

The university has also supported me in bringing my research to the public and allowing me to creatively design curriculum around the topics I research and write about.

The Institute for Research on Women was a nurturing ground for the ideas in this book, and so I thank the institute and all the members of the 2021 Futures Seminar. I want to give a special shout-out to directors Arlene Stein and Sarah Tobias, and fellows Ashley Clemons, Sally Goldfarb, Haylee Harrell, Carter Mathes, and Mònica Tomàs.

This book would not have been possible without the creativity and tenacity of Marilyn Baffoe-Bonnie. I am so grateful that she came into my life when she did! I look forward to learning from her research and seeing her work hit the stands. I am humbled to be a part of her story and grateful to have her a part of mine.

I dedicate this book to my mom, Liza. She has been there since the beginning, and she has helped me plant every seed. I am exceedingly grateful to her for all she has done for me and all she continues to do for my family.

I am also filled with love and light every time I think of my Woodbury family: Barbara, Hanni, Patrick, and Sam. They too are in every word and every thought of this book. I must also give a special thanks to Sam Woodbury for being my brother. He inspires me and keeps me laughing always.

The last thanks are, of course, the biggest of all! I thank my soulmate Nick Woodbury and our littles, Luca, Mars, and Rafa. Nick is my heart and my brain! His love has taught me the true meaning of intelligence. I love my BlissWood Boys with all of me.

Notes

CHAPTER 1: THINKING INTELLIGENCE

1. I. Zwir, C. Del-Val, Mirka Hintsanen, K. M. Cloninger, R. Romero-Zaliz, A. Mesa, J. Arnedo et al., "Evolution of Genetic Networks for Human Creativity," *Molecular Psychiatry* 27, no. 1 (2022): 354–76.
2. Zwir et al.
3. Jay L. Zagorsky, "Do You Have to Be Smart to Be Rich? The Impact of IQ on Wealth, Income and Financial Distress," *Intelligence* 35, no. 5 (2007): 489–501.
4. Carolyn Tiebout and Norman C. Meier, "Artistic Ability and General Intelligence," *Psychological Monographs* 48, no. 1 (1936): 95.
5. Ian J. Deary, Steve Strand, Pauline Smith, and Cres Fernandes, "Intelligence and Educational Achievement," *Intelligence* 35, no. 1 (2007): 13–21.
6. Thomas D. Castle, "The Relationship of Extracurricular Activity Involvement to IQ, Academic Achievement, Attendance, and Discipline Referrals at a Selected Midwestern High School," Drake University, 1986.
7. Panuwun Joko Nurcahyo, Kusnandar Kusnandar, Didik Rilastiyo Budi, Arfin Deri Listiandi, Henie Kurniawati, and Rindha Widyaningsih, "Does Physical Fitness Correlate with IQ? A Study among Football Student Athletes," *Jurnal Pendidikan Jasmani Dan Olahraga* 6, no. 2 (2021).

8. Robert B. McCall, "Childhood IQ's as Predictors of Adult Educational and Occupational Status," *Science* 197, no. 4302 (1977): 482–83.
9. James D. Roff and Raymound Knight, "Preschizophrenics: Low IQ and Aggressive Symptoms as Predictors of Adult Outcome and Marital Status," *Journal of Nervous and Mental Disease* 168 (1980): 129–32.
10. Baoguo Shi, Lijing Wang, Jiahui Yang, Mengpin Zhang, and Li Xu, "Relationship between Divergent Thinking and Intelligence: An Empirical Study of the Threshold Hypothesis with Chinese Children," *Frontiers in Psychology* 8 (2017): 254.
11. Raphael Woolf, "Plato and the Norms of Thought," *Mind* 122, no. 485 (2013): 171–216.
12. Woolf.
13. Plato, *Plato in Twelve Volumes*, vols. 5 and 6, translated by Paul Shorey (Cambridge, MA: Harvard University Press, 1977).
14. Carolina Kuepper-Tetzel, "Bad Memory? Try the Techniques of the Ancient Greeks," The Learning Scientists, February 2, 2017, https://www.learningscientists.org/blog/2017/2/2-1.
15. Charles H. Calisher, "Taxonomy: What's in a Name? Doesn't a Rose by Any Other Name Smell as Sweet?" *Croatian Medical Journal* 48, no. 2 (2007): 268.
16. Pauline Kleingeld, "Kant's Second Thoughts on Race," *Philosophical Quarterly* 57, no. 229 (2007): 573–92.
17. Mark Larrimore, "Antinomies of Race: Diversity and Destiny in Kant," *Patterns of Prejudice* 42, nos. 4–5 (2008): 341–63.
18. Larrimore.
19. John P. Jackson and Nadine M. Weidman, *Race, Racism, and Science: Social Impact and Interaction* (Santa Barbara, CA: ABC-CLIO, 2004).

20. Felix Waldmann, "David Hume Was a Brilliant Philosopher but Also a Racist Involved in Slavery," *The Scotsman*, July 17, 2020.
21. Stephen G. Alter, "Race, Language, and Mental Evolution in Darwin's Descent of Man," *Journal of the History of the Behavioral Sciences* 43, no. 3 (2007): 239–255.
22. Charles Darwin, *The Descent of Man* (New York: Appleton, 1871).
23. Alter, "Race, Language, and Mental Evolution in Darwin's Descent of Man."
24. Herbert Spencer, *The Principles of Biology*, vol. 1 (Outlook Verlag, 2020).
25. Darwin, *The Descent of Man*.
26. Nicholas W. Gillham, "Sir Francis Galton and the Birth of Eugenics," *Annual Review of Genetics* 35, no. 1 (2001): 83–101.
27. Francis Galton, *Hereditary Genius* (New York: Appleton, 1891).
28. Edwin Black, *War against the Weak: Eugenics and America's Campaign to Create a Master Race* (Washington, DC: Dialog Press, 2012).
29. Black.
30. Charles Benedict Davenport, *Heredity in Relation to Eugenics* (New York: Henry Holt, 1911).
31. Paul A. Lombardo, *Three Generations, No Imbeciles: Eugenics, the Supreme Court, and Buck v. Bell* (Baltimore: Johns Hopkins University Press, 2022).
32. Raymond E. Fancher, "Francis Galton's African Ethnography and Its Role in the Development of His Psychology," *British Journal for the History of Science* 16, no. 1 (1983): 67–79.
33. Edmund Ramsden, "Carving Up Population Science: Eugenics, Demography and the Controversy over the 'Biological Law'of

Population Growth," *Social Studies of Science* 32, nos. 5–6 (2002): 857–99.

34. Laura L. Lovett, "'Fitter Families for Future Firesides': Florence Sherbon and Popular Eugenics," *Public Historian* 29, no. 3 (2007): 69–85.
35. Alexandra Minna Stern, "Making Better Babies: Public Health and Race Betterment in Indiana, 1920–1935," *American Journal of Public Health* 92, no. 5 (2002): 742–52.
36. Ajitha Reddy, "The Eugenic Origins of IQ Testing: Implications for post-Atkins Litigation," *DePaul Law Review* 57 (2007): 667.
37. Dennis Garlick, "Understanding the Nature of the General Factor of Intelligence: The Role of Individual Differences in Neural Plasticity as an Explanatory Mechanism," *Psychological Review* 109, no. 1 (2002): 116.
38. Howard E. Gardner, *Frames of Mind: The Theory of Multiple Intelligences* (New York: Basic Books, 2011).
39. John G. Geake and Peter C. Hansen, "Neural Correlates of Intelligence as Revealed by fMRI of Fluid Analogies," *NeuroImage* 26, no. 2 (2005): 555–64.
40. Arthur W. Toga and Paul M. Thompson, "Genetics of Brain Structure and Intelligence," *Annual Review of Neuroscience* 28, no. 1 (2005): 1–23.

CHAPTER 2: UNDERSTANDING IQ

1. Michael Bulmer, *Francis Galton: Pioneer of Heredity and Biometry* (Baltimore: Johns Hopkins University Press, 2003).
2. Adrian Desmond and James R. Moore, *Darwin* (London: Penguin, 1992).
3. Cera R. Lawrence, "Francis Galton (1822–1911)," *Embryo Project Encyclopedia* (2012).

4. Francis Galton, *Hereditary Genius* (New York: Appleton, 1891).
5. Francis Galton, *Typical Laws of Heredity* (London: William Clowes & Sons, 1877).
6. J. S. Jones, "The Galton Laboratory, University College London," in *Sir Francis Galton, FRS* (London: Palgrave Macmillan, 1993), 190–94.
7. Aida Roige, "Intelligence and IQ Testing," *Eugenics Archives*, April 29, 2014.
8. Francis Galton, *Memories of My Life* (London: Routledge, 2015).
9. Michael Bulmer, "The Development of Francis Galton's Ideas on the Mechanism of Heredity," *Journal of the History of Biology* 32, no. 2 (1999): 263–92.
10. Nicholas Wright Gillham, *A Life of Sir Francis Galton: From African Exploration to the Birth of Eugenics* (Oxford: Oxford University Press, 2001).
11. Tahir Pervez and F. S. Kazmi, "Paradigm of Intelligence as a Factor of Political Control," *Cultural-Historical Psychology* 7, no. 2 (2011): 60–65.
12. Alan S. Kaufman, *IQ Testing 101* (New York: Springer, 2009).
13. Thomas J. Hally, "A Brief History of IQ Tests," *Pridobljeno* 15, no. 5 (2015): 2018.
14. Serge Nicolas, Bernard Andrieu, Jean-Claude Croizet, Rasyid B. Sanitioso, and Jeremy Trevelyan Burman, "Sick? Or Slow? On the Origins of Intelligence as a Psychological Object," *Intelligence* 41, no. 5 (2013): 699–711.
15. L. T. Benjamin, "The Birth of American Intelligence Testing," *Monitor on Psychology* 40, no. 1 (2009): 20.
16. James Trent, *Inventing the Feeble Mind: A History of Intellectual Disability in the United States* (New York: Oxford University Press, 2016).

17. Natalie Frank, "Intelligence Testing and the Beginning of Eugenics," *Humanities* (2022).
18. Carl Zimmer, *She Has Her Mother's Laugh: The Powers, Perversions, and Potential of Heredity* (New York: Dutton, 2019).
19. Kaufman, *IQ Testing 101*.
20. Carl Campbell Brigham, *A Study of American Intelligence* (Princeton, NJ: Princeton University Press, 1922).
21. Stephen Jay Gould, *The Mismeasure of Man* (New York: Norton, 1996).
22. Lyndon B. Johnson, "President Lyndon B. Johnson's Remarks at the Signing of the Immigration Bill, Liberty Island, New York, October 3, 1965," Lyndon Baines Johnson Library and Museum, http://www.lbjlib.utexas.edu/johnson/archives.hom/speeches.hom/651003.asp.
23. Ajitha Reddy, "The Eugenic Origins of IQ Testing: Implications for Post-Atkins Litigation," *DePaul Law Review* 57 (2007): 667.
24. Richard J. Evans, *The Third Reich in Power*, vol. 2 (New York: Penguin, 2006).
25. Jonathan C. Friedman, ed., *The Routledge History of the Holocaust* (Hoboken, NJ: Taylor & Francis, 2010).
26. Mitchell Leslie, "The Vexing Legacy of Lewis Terman," *Stanford Magazine*, July/August 2000.
27. Kaufman, *IQ Testing 101*.
28. Kaufman.
29. William T. Dickens and James R. Flynn, "Black Americans Reduce the Racial IQ Gap: Evidence from Standardization Samples," *Psychological Science* 17, no. 10 (2006): 913–20.
30. Dickens and Flynn.
31. Joel N. Shurkin, *Broken Genius: The Rise and Fall of William Shockley, Creator of the Electronic Age* (London: Palgrave Macmillan, 2006).

32. John P. Jackson Jr., "Arthur Jensen, Evolutionary Biology, and Racism," *History of Psychology* (2022).
33. Richard C. Lewontin and Richard Levins, "Stephen Jay Gould—What Does It Mean to Be a Radical?" *Monthly Review* 54, no. 6 (2002): 17.
34. Jean Collins, "Open Letter to a Revolutionary," *Black Voice,* April 27 1973, https://search.library.wisc.edu/digital/ATKMZ3SXFXIPI583/pages/AQ7GSRBOL7DHVF8B?as=text&view=one.
35. Gould, *The Mismeasure of Man.*
36. Steven Rose, Richard Charles Lewontin, and Leon Kamin, "Not in Our Genes: Biology, Ideology and Human Nature," *Wilson Quarterly* 152 (1984).
37. Isaac L. Woods and Scott L. Graves, "The Fortieth Anniversary of Larry PV Riles: Cognitive Assessment and Black Children," *Contemporary School Psychology* 25, no. 2 (2021): 137–39.
38. Gerald Markowitz and David Rosner, *Lead Wars: The Politics of Science and the Fate of America's Children,* vol. 24 (Berkeley: University of California Press, 2014).
39. Marguerite Holloway, "Flynn's Effect," *Scientific American* 280, no. 1 (1999): 37–38.
40. James R. Flynn, "Searching for Justice: The Discovery of IQ Gains over Time," *American Psychologist* 54, no. 1 (1999): 5.
41. Ulric Neisser, Gwyneth Boodoo, Thomas J. Bouchard Jr., A. Wade Boykin, Nathan Brody, Stephen J. Ceci, Diane F. Halpern et al., "Intelligence: Knowns and Unknowns," *American Psychologist* 51, no. 2 (1996): 77.
42. Richard J. Herrnstein and Charles Murray, *The Bell Curve: Intelligence and Class Structure in American Life* (New York: Simon & Schuster, 2010).

43. Parul Sehgal, "Charles Murray Returns, Nodding to Caution but Still Courting Controversy," *New York Times*, February 12, 2020, https://www.nytimes.com/2020/02/12/books/review-human-diversity-charles-murray.html.
44. Linda S. Gottfredson, "Mainstream Science on Intelligence: An Editorial with 52 Signatories, History, and Bibliography," *Intelligence* 24, no. 1 (1997): 13–23.
45. Neisser et al., "Intelligence: Knowns and Unknowns."
46. Radiolab, *G: Problem Space*, podcast, June 14, 2019, https://radiolab.org/episodes/g-problem-space.
47. Ulric Neisser, "Rising Scores on Intelligence Tests: Test Scores Are Certainly Going up All Over the World, but Whether Intelligence Itself Has Risen Remains Controversial," *American Scientist* 85, no. 5 (1997): 440–47.
48. Scott B. Kaufman, "What Do IQ Tests Test? Interview with Psychologist W. Joel Schneider," *Scientific American*, February 3, 2014, https://blogs.scientificamerican.com/beautiful-minds/what-do-iq-tests-test-interview-with-psychologist-w-joel-schneider/.
49. Kaufman.
50. Anett Nyaradi, Jianghong Li, Siobhan Hickling, Jonathan Foster, and Wendy H. Oddy, "The Role of Nutrition in Children's Neurocognitive Development, from Pregnancy through Childhood," *Frontiers in Human Neuroscience* 7 (2013): 97.
51. Stephen J. Schoenthaler, Ian D. Bier, Kelly Young, Dennis Nichols, and Susan Jansenns, "The Effect of Vitamin-Mineral Supplementation on the Intelligence of American Schoolchildren: A Randomized, Double-Blind Placebo-Controlled Trial," *Journal of Alternative and Complementary Medicine* 6, no. 1 (2000): 19–29.

52. Richard E. Nisbett, *Intelligence and How to Get It: Why Schools and Cultures Count* (New York: Norton, 2009).
53. Angela Lee Duckworth, Patrick D. Quinn, Donald R. Lynam, Rolf Loeber, and Magda Stouthamer-Loeber, "Role of Test Motivation in Intelligence Testing," *Proceedings of the National Academy of Sciences* 108, no. 19 (2011): 7716–20.
54. Adam L. Alter, Joshua Aronson, John M. Darley, Cordaro Rodriguez, and Diane N. Ruble, "Rising to the Threat: Reducing Stereotype Threat by Reframing the Threat as a Challenge," *Journal of Experimental Social Psychology* 46, no. 1 (2010): 166–71.
55. Brian Spitzer and Joshua Aronson, "Minding and Mending the Gap: Social Psychological Interventions to Reduce Educational Disparities," *British Journal of Educational Psychology* 85, no. 1 (2015): 1–18.
56. Dorothy Roberts, *Fatal Invention: How Science, Politics, and Big Business Re-create Race in the Twenty-First Century* (New York: New Press/ORIM, 2011).

CHAPTER 3: THE NATURE OF INTELLIGENCE

1. Chris Gunter, "Single Nucleotide Polymorphisms (Snps)," National Human Genome Research Institute, 2022, https://www.genome.gov/genetics-glossary/Single-Nucleotide-Polymorphisms.
2. Gunter.
3. Mengjin Zhu and Shuhong Zhao, "Candidate Gene Identification Approach: Progress and Challenges," *International Journal of Biological Sciences* 3, no. 7 (2007): 420.
4. Tom Strachan and A. P. Read, *Human Molecular Genetics*, Garland Science," *Edition, Kapitel* 13 (2011): 418.
5. Strachan and Read.

6. Teri A. Manolio, "Genomewide Association Studies and Assessment of the Risk of Disease," *New England Journal of Medicine* 363, no. 2 (2010): 166–76.
7. William S. Bush and Jason H. Moore, "Chapter 11: Genome-wide Association Studies," *PLoS Computational Biology* 8, no. 12 (2012): e1002822.
8. https://www.ncbi.nlm.nih.gov/pmc/articles/PMC5501872/
9. Robert Plomin and Sophie Von Stumm, "The New Genetics of Intelligence," *Nature Reviews Genetics* 19, no. 3 (2018): 148–59.
10. Delilah Zabaneh, Eva Krapohl, H. A. Gaspar, Charles Curtis, S. Hong Lee, Hamel Patel, Stephen Newhouse et al., "A Genome-Wide Association Study for Extremely High Intelligence," *Molecular Psychiatry* 23, no. 5 (2018): 1226–32.
11. Evan Charney, "Genes, Behavior, and Behavior Genetics," *Wiley Interdisciplinary Reviews: Cognitive Science* 8, nos. 1–2 (2017): e1405.
12. Kyung Hee Kim and Darya Zabelina, "Cultural Bias in Assessment: Can Creativity Assessment Help?" *International Journal of Critical Pedagogy* 6, no. 2 (2015).
13. Maureen G. Maguire, Gui-shuang Ying, Glenn J. Jaffe, Cynthia A. Toth, Ebenezer Daniel, Juan Grunwald, Daniel F. Martin, Stephanie A. Hagstrom, and CATT Research Group, "Single-Nucleotide Polymorphisms Associated with Age-Related Macular Degeneration and Lesion Phenotypes in the Comparison of Age-Related Macular Degeneration Treatments Trials," *JAMA Ophthalmology* 134, no. 6 (2016): 674–81.
14. Puya Gharahkhani, Eric Jorgenson, Pirro Hysi, Anthony P. Khawaja, Sarah Pendergrass, Xikun Han, Jue Sheng Ong et al., "Genome-Wide Meta-analysis Identifies 127 Open-Angle Glaucoma Loci with Consistent Effect across Ancestries," *Nature Communications* 12, no. 1 (2021): 1–16.

15. Carina Törn, David Hadley, Hye-Seung Lee, William Hagopian, Åke Lernmark, Olli Simell, Marian Rewers et al., "Role of Type 1 Diabetes–Associated SNPs on Risk of Autoantibody Positivity in the TEDDY Study," *Diabetes* 64, no. 5 (2015): 1818–29.
16. Aysu Okbay, Yeda Wu, Nancy Wang, Hariharan Jayashankar, Michael Bennett, Seyed Moeen Nehzati, Julia Sidorenko et al., "Polygenic Prediction of Educational Attainment within and between Families from Genome-Wide Association Analyses in 3 Million Individuals," *Nature Genetics* 54, no. 4 (2022): 437–49.
17. Giorgio Sirugo, Scott M. Williams, and Sarah A. Tishkoff, "The Missing Diversity in Human Genetic Studies," *Cell* 177, no. 1 (2019): 26–31.
18. William T. Dickens and James R. Flynn, "Black Americans Reduce the Racial IQ Gap: Evidence from Standardization Samples," *Psychological Science* 17, no. 10 (2006): 913–20.
19. Marcus W. Feldman and S. P. Otto, "Twin Studies, Heritability, and Intelligence," *Science* 278, no. 5342 (1997): 1383–87.
20. Michael Cummings, *Human Heredity: Principles and Issues* (Boston: Cengage Learning, 2015).
21. Richard D. Rende, Robert Plomin, and Steven G. Vandenberg, "Who Discovered the Twin Method?" *Behavior Genetics* 20, no. 2 (1990): 277–85.
22. Ronald Fletcher, *Science, Ideology, and the Media: The Cyril Burt Scandal* (London: Routledge, 2017).
23. Malcolm Pines, "The Development of the Psychodynamic Movement," in German Berrios and Hugh Freeman, eds., *150 Years of British Psychiatry, 1841–1991* (London: Gaskell, 1991).
24. Leslie Spenser Hearnshaw, *Cyril Burt: Psychologist* (New York: Vintage, 1979).

25. Sylia Wilson, Kevin Haroian, William G. Iacono, Robert F. Krueger, James J. Lee, Monica Luciana, Stephen M. Malone, Matt McGue, Glenn I. Roisman, and Scott Vrieze, "Minnesota Center for Twin and Family Research," *Twin Research and Human Genetics* 22, no. 6 (2019): 746–52.
26. Ian J. Deary, Frank M. Spinath, and Timothy C. Bates, "Genetics of Intelligence," *European Journal of Human Genetics* 14 (2006): 690–700.
27. Paul Lichtenstein, Patrick F. Sullivan, Sven Cnattingius, Margaret Gatz, Sofie Johansson, Eva Carlström, Camilla Björk et al., "The Swedish Twin Registry in the Third Millennium: An Update," *Twin Research and Human Genetics* 9, no. 6 (2006): 875–82.
28. Jay Joseph, *The Trouble with Twin Studies* (New York: Routledge, 2014).
29. Jay Joseph, "Twenty-Two Invalidating Aspects of the Minnesota Study of Twins Reared Apart (MISTRA)," published online, 2018.
30. Jay Joseph, "A Reevaluation of the 1990 'Minnesota Study of Twins Reared Apart' IQ Study," *Human Development* 66, no. 1 (2022): 48–65.
31. Jay Joseph.
32. Adam Miller, "The Pioneer Fund: Bankrolling the Professors of Hate," *Journal of Blacks in Higher Education* 6 (1994): 58–61.
33. Carl E. G. Bruder, Arkadiusz Piotrowski, Antoinet A. C. J. Gijsbers, Robin Andersson, Stephen Erickson, Teresita Diaz de Ståhl, Uwe Menzel et al., "Phenotypically Concordant and Discordant Monozygotic Twins Display Different DNA Copy-Number-Variation Profiles," *American Journal of Human Genetics* 82, no. 3 (2008): 763–71.

34. National Academies of Sciences, Engineering, and Medicine, *Human Genome Editing: Science, Ethics, and Governance* (Washington, DC: National Academies Press, 2017).
35. Vera Lucia Raposo, "The First Chinese Edited Babies: A Leap of Faith in Science," *JBRA Assisted Reproduction* 23, no. 3 (2019): 197.

CHAPTER 4: NURTURING INTELLIGENCE

1. Nancy L. Segal and Yesika S. Montoya, *Accidental Brothers: The Story of Twins Exchanged at Birth and the Power of Nature and Nurture* (New York: St. Martin's Press, 2018).
2. Segal and Montoya.
3. Erika Hayasaki, "Identical Twins Hint at How Environments Change Gene Expression," *The Atlantic*, 2018.
4. Scott M. Langevin and Karl T. Kelsey, "The Fate Is Not Always Written in the Genes: Epigenomics in Epidemiologic Studies," *Environmental and Molecular Mutagenesis* 54, no. 7 (2013): 533–41.
5. Jolie D. Barter and Thomas C. Foster, "Aging in the Brain: New Roles of Epigenetics in Cognitive Decline," *Neuroscientist* 24, no. 5 (2018): 516–25.
6. Barter and Foster.
7. "Epigenome," National Human Genome Research Institute, 2022, https://www.genome.gov/genetics-glossary/Epigenome.
8. Lisa D. Moore, Thuc Le, and Guoping Fan, "DNA Methylation and Its Basic Function," *Neuropsychopharmacology* 38, no. 1 (2013): 23–38.
9. B. Alaskhar Alhamwe, R. Khalaila, J. Wolf, V. von Bülow, H. Harb, F. Alhamdan, C. S. Hii, S. L. Prescott, A. Ferrante, H. Renz, and H. Garn, "Histone Modifications and Their Role in Epigenetics of Atopy and Allergic Diseases,"

Allergy, Asthma & Clinical Immunology 14, no. 1 (2018): 1–16.

10. Irene Lacal and Rossella Ventura, "Epigenetic Inheritance: Concepts, Mechanisms and Perspectives," *Frontiers in Molecular Neuroscience* (2018): 292.
11. Michela Fagiolini, Catherine L. Jensen, and Frances A. Champagne, "Epigenetic Influences on Brain Development and Plasticity," *Current Opinion in Neurobiology* 19, no. 2 (2009): 207–12.
12. Bastiaan T. Heijmans, Elmar W. Tobi, Aryeh D. Stein, Hein Putter, Gerard J. Blauw, Ezra S. Susser, P. Eline Slagboom, and L. H. Lumey, "Persistent Epigenetic Differences Associated with Prenatal Exposure to Famine in Humans," *Proceedings of the National Academy of Sciences* 105, no. 44 (2008): 17046–49.
13. Natan P. F. Kellermann, "Epigenetic Transmission of Holocaust Trauma: Can Nightmares Be Inherited," *Israel Journal of Psychiatry and Related Sciences* 50, no. 1 (2013): 33–39.
14. Jenny Hsieh and Xinyu Zhao, "Genetics and Epigenetics in Adult Neurogenesis," *Cold Spring Harbor Perspectives in Biology* 8, no. 6 (2016): a018911.
15. María Camila Cortés-Albornoz, Danna Paola García-Guáqueta, Alberto Velez-van-Meerbeke, and Claudia Talero-Gutiérrez, "Maternal Nutrition and Neurodevelopment: A Scoping Review," *Nutrients* 13, no. 10 (2021): 3530.
16. Pauline Dimofski, David Meyre, Natacha Dreumont, and Brigitte Leininger-Muller, "Consequences of Paternal Nutrition on Offspring Health and Disease," *Nutrients* 13, no. 8 (2021): 2818.
17. Daniel A. Notterman and Colter Mitchell, "Epigenetics and Understanding the Impact of Social Determinants of Health," *Pediatric Clinics* 62, no. 5 (2015): 1227–40.

18. Jennifer C. Chan, Bridget M. Nugent, and Tracy L. Bale, "Parental Advisory: Maternal and Paternal Stress Can Impact Offspring Neurodevelopment," *Biological Psychiatry* 83, no. 10 (2018): 886–94.
19. Jakob A. Kaminski, Florian Schlagenhauf, Michael Rapp, Swapnil Awasthi, Barbara Ruggeri, Lorenz Deserno, Tobias Banaschewski et al., "Epigenetic Variance in Dopamine D2 Receptor: A Marker of IQ Malleability?" *Translational Psychiatry* 8, no. 1 (2018): 1–11.
20. Kylie Garber Bezdek and Eva H. Telzer, "Have No Fear, the Brain Is Here! How Your Brain Responds to Stress," *Frontiers for Young Minds* 5, no. 71 (December 2017): 1–8.
21. Bezdek and Telzer.
22. Habib Yaribeygi, Yunes Panahi, Hedayat Sahraei, Thomas P. Johnston, and Amirhossein Sahebkar, "The Impact of Stress on Body Function: A Review," *EXCLI Journal* 16 (2017): 1057.
23. Yaribeygi et al.
24. Susanne Vogel and Lars Schwabe, "Learning and Memory under Stress: Implications for the Classroom," *NPJ Science of Learning* 1, no. 1 (2016): 1–10.
25. Yaribeygi et al., "The Impact of Stress on Body Function."
26. Huan Song, Johanna Sieurin, Karin Wirdefeldt, Nancy L. Pedersen, Catarina Almqvist, Henrik Larsson, Unnur A. Valdimarsdóttir, and Fang Fang, "Association of Stress-Related Disorders with Subsequent Neurodegenerative Diseases," *JAMA Neurology* 77, no. 6 (2020): 700–709.
27. Song et al.
28. Song et al.
29. Brian S. Mohlenhoff, Aoife O'Donovan, Michael W. Weiner, and Thomas C. Neylan, "Dementia Risk in Posttraumatic Stress Disorder: The Relevance of Sleep-Related Abnormalities

in Brain Structure, Amyloid, and Inflammation," *Current Psychiatry Reports* 19, no. 11 (2017): 1–9.

30. Viviana J. Mancilla, Noah C. Peeri, Talisa Silzer, Riyaz Basha, Martha Felini, Harlan P. Jones, Nicole Phillips, Meng-Hua Tao, Srikantha Thyagarajan, and Jamboor K. Vishwanatha, "Understanding the Interplay between Health Disparities and Epigenomics," *Frontiers in Genetics* 11 (2020): 903.

CHAPTER 5: THE GROWTH MINDSET

1. Elena P. Moreno-Jiménez, Miguel Flor-García, Julia Terreros-Roncal, Alberto Rábano, Fabio Cafini, Noemí Pallas-Bazarra, Jesús Ávila, and María Llorens-Martín, "Adult Hippocampal Neurogenesis Is Abundant in Neurologically Healthy Subjects and Drops Sharply in Patients with Alzheimer's Disease," *Nature Medicine* 25, no. 4 (2019): 554–60.
2. Chiara F. Tagliabue, Sara Assecondi, Giulia Cristoforetti, and Veronica Mazza, "Learning by Task Repetition Enhances Object Individuation and Memorization in the Elderly," *Scientific Reports* 10, no. 1 (2020): 1–12.
3. Ícaro J. S. Ribeiro, Rafael Pereira, Ivna V. Freire, Bruno G. de Oliveira, Cezar A. Casotti, and Eduardo N. Boery, "Stress and Quality of Life among University Students: A Systematic Literature Review," *Health Professions Education* 4, no. 2 (2018): 70–77.
4. Richa Burman and Tulsee Giri Goswami, "A Systematic Literature Review of Work Stress," *International Journal of Management Studies* 3, no. 9 (2018): 112–32.
5. William Stixrud and Ned Johnson, *The Self-Driven Child: The Science and Sense of Giving Your Kids More Control over Their Lives* (New York: Penguin, 2019).

6. "Silicon Valley's Workforce Is Feeling More Burned Out than before the Pandemic, with Nearly 70% Reporting Work-from-Home Exhaustion," *Business Insider México*, 2020, https://businessinsider.mx/silicon-valleys-workforce-is-feeling-more-burned-out-than-before-the-pandemic-with-nearly-70-reporting-work-home-home-exhaustion/.
7. Stixrud and Johnson, *The Self-Driven Child*.
8. Suniya S. Luthar, Phillip J. Small, and Lucia Ciciolla, "Adolescents from Upper Middle Class Communities: Substance Misuse and Addiction across Early Adulthood," *Development and Psychopathology* 30, no. 2 (2018): 715–16.
9. Ray Hart, Michael Casserly, Renata Uzzell, Moses Palacios, Amanda Corcoran, and Liz Spurgeon, "Student Testing in America's Great City Schools: An Inventory and Preliminary Analysis," Council of the Great City Schools, 2015.
10. Betsy Ng, "The Neuroscience of Growth Mindset and Intrinsic Motivation," *Brain Sciences* 8, no. 2 (2018): 20.
11. Carol Dweck, "Carol Dweck Revisits the Growth Mindset," *Education Week* 35, no. 5 (2015): 20–24.
12. Hans S. Schroder, Tim P. Moran, M. Brent Donnellan, and Jason S. Moser, "Mindset Induction Effects on Cognitive Control: A Neurobehavioral Investigation," *Biological Psychology* 103 (2014): 27–37.
13. Yuchen Song, Michael M. Barger, and Kristen L. Bub, "The Association between Parents' Growth Mindset and Children's Persistence and Academic Skills," in *Frontiers in Education* (2022), 525.
14. Joann Deak, *Your Fantastic Elastic Brain* (Naperville, IL: Little Pickle Press, 2010).
15. Jérémie Blanchette Sarrasin, Lucian Nenciovici, Lorie-Marlène Brault Foisy, Geneviève Allaire-Duquette, Martin

Riopel, and Steve Masson, "Effects of Teaching the Concept of Neuroplasticity to Induce a Growth Mindset on Motivation, Achievement, and Brain Activity: A Meta-analysis," *Trends in Neuroscience and Education* 12 (2018): 22–31.

16. Keith Heggart, "Developing a Growth Mindset in Teachers and Staff," Edutopia, February 4, 2015.
17. Heggart.
18. Laurie Murphy and Lynda Thomas, "Dangers of a Fixed Mindset: Implications of Self-Theories Research for Computer Science Education," in *Proceedings of the 13th Annual Conference on Innovation and Technology in Computer Science Education* (2008): 271–75.
19. Mari Rege, Paul Hanselman, Ingeborg Foldøy Solli, Carol S. Dweck, Sten Ludvigsen, Eric Bettinger, Robert Crosnoe et al., "How Can We Inspire Nations of Learners? An Investigation of Growth Mindset and Challenge-Seeking in Two Countries," *American Psychologist* 76, no. 5 (2021): 755.
20. Kyla Haimovitz and Carol S. Dweck, "What Predicts Children's Fixed and Growth Intelligence Mind-sets? Not Their Parents' Views of Intelligence but Their Parents' Views of Failure," *Psychological Science* 27, no. 6 (2016): 859–69.
21. Chang Seek Lee, Sun Ui Park, and Yeoun Kyoung Hwang, "The Structural Relationship between Mother's Parenting Stress and Child's Well-being: The Mediating Effects of Mother's Growth Mindset and Hope," *Indian Journal of Science and Technology* 9, no. 36 (2016): 1–6.
22. Mary Alice Barksdale-Ladd and Karen F. Thomas, "What's at Stake in High-Stakes Testing: Teachers and Parents Speak Out," *Journal of Teacher Education* 51, no. 5 (2000): 384–97.
23. Lisa S. Blackwell, Kali H. Trzesniewski, and Carol Sorich Dweck, "Implicit Theories of Intelligence Predict Achievement

across an Adolescent Transition: A Longitudinal Study and an Intervention," *Child Development* 78, no. 1 (2007): 246–63.
24. Anindito Aditomo, "Students' Response to Academic Setback: 'Growth Mindset' as a Buffer against Demotivation," *International Journal of Educational Psychology* 4, no. 2 (2015): 198–222.
25. Alireza Yousefy and Maryam Gordanshekan, "A Review on Development of Self-Directed Learning," *Iranian Journal of Medical Education* 10, no. 5 (2011).
26. David S. Yeager, Paul Hanselman, Gregory M. Walton, Jared S. Murray, Robert Crosnoe, Chandra Muller, Elizabeth Tipton et al., "A National Experiment Reveals Where a Growth Mindset Improves Achievement," *Nature* 573, no. 7774 (2019): 364–69.
27. Soo Jeoung Han and Vicki Stieha, "Growth Mindset for Human Resource Development: A Scoping Review of the Literature with Recommended Interventions," *Human Resource Development Review* 19, no. 3 (2020): 309–31.
28. Blackwell, Trzesniewski, and Dweck, "Implicit Theories of Intelligence Predict Achievement across an Adolescent Transition."
29. Eleanor O'Rourke, Kyla Haimovitz, Christy Ballweber, Carol Dweck, and Zoran Popović, "Brain Points: A Growth Mindset Incentive Structure Boosts Persistence in an Educational Game," in *Proceedings of the SIGCHI Conference on Human Factors in Computing Systems* (2014): 3339–48.
30. Thomas Sullivan and Nadine Page, "A Competency Based Approach to Leadership Development: Growth Mindset in the Workplace," in *New Leadership in Strategy and Communication* (Cham, Switzerland: Springer, Cham, 2020), 179–89.
31. Herminia Ibarra, Aneeta Rattan, and Anna Johnston, "Satya Nadella at Microsoft: Instilling a Growth Mindset," *Harvard*

Business Review, case no. LBS128 (2018): 1–22.

32. Blanchette Sarrasin et al., "Effects of Teaching the Concept of Neuroplasticity to Induce a Growth Mindset on Motivation, Achievement, and Brain Activity: A Meta-analysis."
33. Jason S. Moser, Hans S. Schroder, Carrie Heeter, Tim P. Moran, and Yu-Hao Lee, "Mind Your Errors: Evidence for a Neural Mechanism Linking Growth Mind-set to Adaptive Posterror Adjustments," *Psychological Science* 22, no. 12 (2011): 1484–89.
34. Moser et al.
35. Christina Bejjani, Samantha DePasque, and Elizabeth Tricomi, "Intelligence Mindset Shapes Neural Learning Signals and Memory," *Biological Psychology* 146 (2019): 107715.
36. Bejjani, DePasque, and Tricomi.
37. Blanchette Sarrasin et al., "Effects of Teaching the Concept of Neuroplasticity to Induce a Growth Mindset on Motivation, Achievement, and Brain Activity: A Meta-analysis."
38. Hae Yeon Lee, Jeremy P. Jamieson, Adriana S. Miu, Robert A. Josephs, and David S. Yeager, "An Entity Theory of Intelligence Predicts Higher Cortisol Levels When High School Grades Are Declining," *Child Development* 90, no. 6 (2019): e849–e867.
39. Lee et al.
40. Guang Zeng, Hanchao Hou, and Kaiping Peng, "Effect of Growth Mindset on School Engagement and Psychological Well-being of Chinese Primary and Middle School Students: The Mediating Role of Resilience," *Frontiers in Psychology* 7 (2016): 1873.
41. Aditomo, "Students' Response to Academic Setback."
42. Mengting Li, Weiqiao Fan, and Frederick T. L. Leong, "Growth Mindset of Intelligence Reduces Counterproductive Workplace Behavior: A Mediation Analysis of Occupational

Stress," *International Journal of Selection and Assessment* 29, nos. 3–4 (2021): 519–26.

43. Martin Huecker, Jacob Shreffler, and Daniel Danzl, "COVID-19: Optimizing Healthcare Provider Wellness and Posttraumatic Growth," *American Journal of Emergency Medicine* (2020).
44. Hans S. Schroder, Matthew M. Yalch, Sindes Dawood, Courtney P. Callahan, M. Brent Donnellan, and Jason S. Moser, "Growth Mindset of Anxiety Buffers the Link between Stressful Life Events and Psychological Distress and Coping Strategies," *Personality and Individual Differences* 110 (2017): 23–26.

CHAPTER 6: FROM MIND TO MINDFUL

1. Ruth A. Baer, "Mindfulness Training as a Clinical Intervention: A Conceptual and Empirical Review," *Clinical Psychology: Science and Practice* 10, no. 2 (2003): 125.
2. Jon Kabat-Zinn and Thich Nhat Hanh, *Full Catastrophe Living: Using the Wisdom of Your Body and Mind to Face Stress, Pain, and Illness* (New York: Delta, 2009).
3. Bassam Khoury, Manoj Sharma, Sarah E. Rush, and Claude Fournier, "Mindfulness-Based Stress Reduction for Healthy Individuals: A Meta-analysis," *Journal of Psychosomatic Research* 78, no. 6 (2015): 519–28.
4. Darren L. Dunning, Kirsty Griffiths, Willem Kuyken, Catherine Crane, Lucy Foulkes, Jenna Parker, and Tim Dalgleish, "Research Review: The Effects of Mindfulness-Based Interventions on Cognition and Mental Health in Children and Adolescents: A Meta-analysis of Randomized Controlled Trials," *Journal of Child Psychology and Psychiatry* 60, no. 3 (2019): 244–58.

5. Tim Whitfield, Thorsten Barnhofer, Rebecca Acabchuk, Avi Cohen, Michael Lee, Marco Schlosser, Eider M. Arenaza-Urquijo et al., "The Effect of Mindfulness-Based Programs on Cognitive Function in Adults: A Systematic Review and Meta-analysis," *Neuropsychology Review* (2021): 1–26.
6. Emily K. Lindsay, Brian Chin, Carol M. Greco, Shinzen Young, Kirk W. Brown, Aidan G. C. Wright, Joshua M. Smyth, Deanna Burkett, and J. David Creswell, "How Mindfulness Training Promotes Positive Emotions: Dismantling Acceptance Skills Training in Two Randomized Controlled Trials," *Journal of Personality and Social Psychology* 115, no. 6 (2018): 944.
7. Eileen Luders, Florian Kurth, Emeran A. Mayer, Arthur W. Toga, Katherine L. Narr, and Christian Gaser, "The Unique Brain Anatomy of Meditation Practitioners: Alterations in Cortical Gyrification," *Frontiers in Human Neuroscience* 6 (2012): 34.
8. M. B. Cladder-Micus, Joël van Aalderen, A. R. T. Donders, Jan Spijker, J. N. Vrijsen, and A. E. M. Speckens, "Cognitive Reactivity as Outcome and Working Mechanism of Mindfulness-Based Cognitive Therapy for Recurrently Depressed Patients in Remission," *Cognition and Emotion* 32, no. 2 (2018): 371–78.
9. Arielle L. Klopsis, "The Impact of a Single Session of Mindfulness Meditation on the Attentional Blink in Non-Meditators," 2020.
10. Maddalena Boccia, Laura Piccardi, and Paola Guariglia, "The Meditative Mind: A Comprehensive Meta-analysis of MRI Studies," *BioMed Research International* (2015).
11. Boccia, Piccardi, and Guariglia.
12. Boccia, Piccardi, and Guariglia.
13. Whitfield et al., "The Effect of Mindfulness-Based Programs on Cognitive Function in Adults."

14. Whitfield et al.
15. Whitfield et al.
16. Feng Ling Wang, Qian Yun Tang, Lu Lu Zhang, Jing Jing Yang, Yu Li, Hua Peng, and Shu Hong Wang, "Effects of Mindfulness-Based Interventions on Dementia Patients: A Meta-analysis," *Western Journal of Nursing Research* 42, no. 12 (2020): 1163–73.
17. Troy A. Richter and Richard G. Hunter, "Epigenetics in Post-traumatic Stress Disorder," in *Epigenetics in Psychiatry* (San Diego: Academic Press, 2021), 429–50.
18. Shin Hashizume, Masako Nakano, Kenta Kubota, Seiichi Sato, Nobuaki Himuro, Eiji Kobayashi, Akinori Takaoka, and Mineko Fujimiya, "Mindfulness Intervention Improves Cognitive Function in Older Adults by Enhancing the Level of miRNA-29c in Neuron-Derived Extracellular Vesicles," *Scientific Reports* 11, no. 1 (2021): 1–14.
19. Jeffrey R. Bishop, Adam M. Lee, Lauren J. Mills, Paul D. Thuras, Seenae Eum, Doris Clancy, Christopher R. Erbes, Melissa A. Polusny, Gregory J. Lamberty, and Kelvin O. Lim, "Methylation of FKBP5 and SLC6A4 in Relation to Treatment Response to Mindfulness Based Stress Reduction for Posttraumatic Stress Disorder," *Frontiers in Psychiatry* 9 (2018): 418.
20. Concetta Gardi, Teresa Fazia, Blerta Stringa, and Fabio Giommi, "A Short Mindfulness Retreat Can Improve Biological Markers of Stress and Inflammation," *Psychoneuroendocrinology* 135 (2022): 105579.
21. Khoury et al., "Mindfulness-Based Stress Reduction for Healthy Individuals: A Meta-analysis."
22. Boccia, Piccardi, and Guariglia, "The Meditative Mind."
23. Shaji John, Satish Kumar Verma, and Gulshan L. Khanna, "The Effect of Mindfulness Meditation on HPA-Axis in

Pre-competition Stress in Sports Performance of Elite Shooters," *National Journal of Integrated Research in Medicine* 2, no. 3 (2011): 15–21.

24. Gardi et al., "A Short Mindfulness Retreat Can Improve Biological Markers of Stress and Inflammation."
25. Jon Kabat-Zinn, "Some Reflections on the Origins of MBSR, Skillful Means, and the Trouble with Maps," in *Mindfulness* (London: Routledge, 2013), 281–306.
26. Kabat-Zinn.
27. Saki F. Santorelli, Jon Kabat-Zinn, Melissa Blacker, Florence Meleo-Meyer, and Lynn Koerbel, "Mindfulness-Based Stress Reduction (MBSR) Authorized Curriculum Guide," Center for Mindfulness in Medicine, Health Care, and Society (CFM), University of Massachusetts Medical School, 2017.
28. Paul Grossman, Ludger Niemann, Stefan Schmidt, and Harald Walach, "Mindfulness-Based Stress Reduction and Health Benefits: A Meta-analysis," *Journal of Psychosomatic Research* 57, no. 1 (2004): 35–43.
29. John J. Miller, Ken Fletcher, and Jon Kabat-Zinn, "Three-Year Follow-up and Clinical Implications of a Mindfulness Meditation-Based Stress Reduction Intervention in the Treatment of Anxiety Disorders," *General Hospital Psychiatry* 17, no. 3 (1995): 192–200.
30. Gardi et al., "A Short Mindfulness Retreat Can Improve Biological Markers of Stress and Inflammation."
31. Brandon W. Qualls, Emily M. Payton, Laura G. Aikens, and Mary G. Carey, "Mindfulness for Outpatient Oncology Nurses: A Pilot Study," *Holistic Nursing Practice* 36, no. 1 (2022): 28–36.
32. Mark A. Craigie, Clare S. Rees, Ali Marsh, and Paula Nathan, "Mindfulness-Based Cognitive Therapy for Generalized

Anxiety Disorder: A Preliminary Evaluation," *Behavioural and Cognitive Psychotherapy* 36, no. 5 (2008): 553–68.

33. Samaneh Abedini, Mojtaba Habibi, Negar Abedini, Thomas M. Achenbach, and Randye J. Semple, "A Randomized Clinical Trial of a Modified Mindfulness-Based Cognitive Therapy for Children Hospitalized with Cancer," *Mindfulness* 12, no. 1 (2021): 141–51.
34. Paul Chadwick, Tracey Newell, and Chas Skinner, "Mindfulness Groups in Palliative Care: A Pilot Qualitative Study," *Spirituality and Health International* 9, no. 3 (2008): 135–44.
35. Kathryn Birnie, Sheila N. Garland, and Linda E. Carlson, "Psychological Benefits for Cancer Patients and Their Partners Participating in Mindfulness-Based Stress Reduction (MBSR)," *Psycho-oncology* 19, no. 9 (2010): 1004–9.
36. Meghal Gagrani, Muneeb A. Faiq, Talvir Sidhu, Rima Dada, Raj K. Yadav, Ramanjit Sihota, Kanwal P. Kochhar, Rohit Verma, and Tanuj Dada, "Meditation Enhances Brain Oxygenation, Upregulates BDNF and Improves Quality of Life in Patients with Primary Open Angle Glaucoma: A Randomized Controlled Trial," *Restorative Neurology and Neuroscience* 36, no. 6 (2018): 741–53.
37. Anna Lardone, Marianna Liparoti, Pierpaolo Sorrentino, Rosaria Rucco, Francesca Jacini, Arianna Polverino, Roberta Minino et al., "Mindfulness Meditation Is Related to Long-Lasting Changes in Hippocampal Functional Topology during Resting State: A Magnetoencephalography Study," *Neural Plasticity* (2018).
38. Grossman et al., "Mindfulness-Based Stress Reduction and Health Benefits."
39. Guichen Li, Hua Yuan, and Wei Zhang, "The Effects of Mindfulness-Based Stress Reduction for Family Caregivers:

Systematic Review," *Archives of Psychiatric Nursing* 30, no. 2 (2016): 292–99.

40. Daniel Campos, Ausiàs Cebolla, Soledad Quero, Juana Bretón-López, Cristina Botella, Joaquim Soler, Javier García-Campayo, Marcelo Demarzo, and Rosa María Baños, "Meditation and Happiness: Mindfulness and Self-Compassion May Mediate the Meditation–Happiness Relationship," *Personality and Individual Differences* 93 (2016): 80–85.
41. Grossman et al., "Mindfulness-Based Stress Reduction and Health Benefits."
42. Julia K. Hutchinson, Jaci C. Huws, and Dusana Dorjee, "Exploring Experiences of Children in Applying a School-Based Mindfulness Programme to Their Lives," *Journal of Child and Family Studies* 27, no. 12 (2018): 3935–51.
43. Jacqueline M. Smith, Katherine S. Bright, Joel Mader, Jennifer Smith, Arfan Raheen Afzal, Charmaine Patterson, Gina Dimitropolous, and Rachael Crowder, "A Pilot of a Mindfulness Based Stress Reduction Intervention for Female Caregivers of Youth Who Are Experiencing Substance Use Disorders," *Addictive Behaviors* 103 (2020): 106223.
44. Akira S. Gutierrez, Sara B. Krachman, Ethan Scherer, Martin R. West, and John D. Gabrieli, "Mindfulness in the Classroom: Learning from a School-Based Mindfulness Intervention through the Boston Charter Research Collaborative," *Transforming Education* (2019).
45. Gutierrez et al.
46. Theodore C. Masters-Waage, Jared Nai, Jochen Reb, Samantha Sim, Jayanth Narayanan, and Noriko Tan, "Going Far Together by Being Here Now: Mindfulness Increases Cooperation in Negotiations," *Organizational Behavior and Human Decision Processes* 167 (2021): 189–205.

47. Aileen M. Pidgeon and Michelle Keye, "Relationship between Resilience, Mindfulness, and Pyschological Well-being in University Students," *International Journal of Liberal Arts and Social Science* 2, no. 5 (2014): 27–32.
48. Ashley Borders, Mitch Earleywine, and Archana Jajodia, "Could Mindfulness Decrease Anger, Hostility, and Aggression by Decreasing Rumination?" *Aggressive Behavior: Official Journal of the International Society for Research on Aggression* 36, no. 1 (2010): 28–44.
49. A. Kamenetz and M. Knight, "Schools Are Embracing Mindfulness, but Practice Doesn't Always Make Perfect," 2020.
50. Ruben Vonderlin, Miriam Biermann, Martin Bohus, and Lisa Lyssenko, "Mindfulness-Based Programs in the Workplace: A Meta-analysis of Randomized Controlled Trials," *Mindfulness* 11, no. 7 (2020): 1579–98.
51. Qualls et al.,"Mindfulness for Outpatient Oncology Nurses: A Pilot Study."
52. Katrin Micklitz, Geoff Wong, and Jeremy Howick, "Mindfulness-Based Programmes to Reduce Stress and Enhance Well-being at Work: A Realist Review," *BMJ Open* 11, no. 3 (2021): e043525.
53. Alexandra Michel, Christine Bosch, and Miriam Rexroth, "Mindfulness as a Cognitive–Emotional Segmentation Strategy: An Intervention Promoting Work–Life Balance," *Journal of Occupational and Organizational Psychology* 87, no. 4 (2014): 733–54.
54. Andreas Wihler, Ute R. Hülsheger, Jochen Reb, and Jochen I. Menges, "It's So Boring—or Is It? Examining the Role of Mindfulness for Work Performance and Attitudes in Monotonous Jobs," *Journal of Occupational and Organizational Psychology* 95, no. 1 (2022): 131–54.

55. Patricia L. Dobkin and Tom A. Hutchinson, "Teaching Mindfulness in Medical School: Where Are We Now and Where Are We Going?" *Medical Education* 47, no. 8 (2013): 768–79.
56. Vonderlin et al., "Mindfulness-Based Programs in the Workplace."
57. Christian Greiser and Jan-Philipp Martini, "Unleashing the Power of Mindfulness in Corporations," Boston Consulting Group, 2018.
58. Greiser and Martini.
59. Greiser and Martini.
60. Norian A. Caporale-Berkowitz, Brittany P. Boyer, Christopher J. Lyddy, Darren J. Good, Aaron B. Rochlen, and Michael C. Parent, "Search Inside Yourself: Investigating the Effects of a Widely Adopted Mindfulness-at-Work Development Program," *International Journal of Workplace Health Management* (2021).
61. David Gelles, *Mindful Work: How Meditation Is Changing Business from the Inside Out* (Boston: Houghton Mifflin Harcourt, 2015).
62. Nell D. Debevoise, "LinkedIn's Mindfulness Lead Branches Out to Cultivate Inside Out Leadership," *Forbes*, September 16, 2021, https://www.forbes.com/sites/nelldebevoise/2021/09/16/linkedins-mindfulness-lead-branches-out-to-cultivate-inside-out-leadership/?sh=10db824546d4.
63. Catherine Clifford, "CEO Marc Benioff: Why We Have 'Mindfulness Zones' Where Employees Put Away Phones, Clear Their Minds," CNBC, November 5, 2019, https://www.cnbc.com/2019/11/05/salesforce-ceo-marc-benioff-why-we-have-mindfulness-zones.html.

CHAPTER 7: LEARNING TO CONNECT

1. Hamdi Serin, "A Comparison of Teacher-Centered and Student-Centered Approaches in Educational Settings,"

International Journal of Social Sciences & Educational Studies 5, no. 1 (2018): 164–67.

2. James Kelly, "Collaborative Learning: Higher Education, Interdependence, and the Authority of Knowledge by Kenneth Bruffee: A Critical Study," *Journal of the National Collegiate Honors Council*, online archive (2002): 82.
3. Robyn M. Gillies, "Cooperative Learning: Review of Research and Practice," *Australian Journal of Teacher Education* (online) 41, no. 3 (2016): 3954.
4. Jiří Dostál and Jan Gregar, *Inquiry-Based Instruction: Concept, Essence, Importance and Contribution* (Olomouc, Czech Republic: Univerzita Palackého v Olomouci, 2015).
5. Terry Heick, "5 Types of Project-Based Learning Symbolize Its Evolution," http://www.teachthought.com/learning/project-based-learning/5-types-of-project-based-learning-symbolize-its-evolution/.
6. Mizuko Ito, Kris Gutiérrez, Sonia Livingstone, Bill Penuel, Jean Rhodes, Katie Salen, Juliet Schor, Julian Sefton-Green, and S. Craig Watkins, *Connected Learning: An Agenda for Research and Design* (Digital Media and Learning Research Hub, 2013).
7. Ito et al.
8. Ito et al.
9. W. Reid Cornwell and Jonathan R. Cornwell, "Connected Learning: A Framework of Observation, Research and Development to Guide the Reform of Education," 2006, http://tcfir.org/whitepapers/Connected%20Learning%20Framework.pdf.
10. Sergey Gavrilets, "Collective Action and the Collaborative Brain," *Journal of the Royal Society Interface* 12, no. 102 (2015): 20141067.

11. Sara Stillesjö, Linnea Karlsson Wirebring, Micael Andersson, Carina Granberg, Johan Lithner, Bert Jonsson, Lars Nyberg, and Carola Wiklund-Hörnqvist, "Active Math and Grammar Learning Engages Overlapping Brain Networks," *Proceedings of the National Academy of Sciences* 118, no. 46 (2021): e2106520118.
12. M.-H. Sohn et al., "Behavioral equivalence, but not neural equivalence—neural evidence of alternative strategies in mathematical thinking," *Nature Neuroscience* 7, 1193–94 (2004).
13. Stillesjö et al.
14. Christina Hinton, Kurt W. Fischer, and Catherine Glennon, "Mind, Brain, and Education," *Mind* (2012).
15. Artur Czeszumski, Sara Eustergerling, Anne Lang, David Menrath, Michael Gerstenberger, Susanne Schuberth, Felix Schreiber, Zadkiel Zuluaga Rendon, and Peter König, "Hyperscanning: A Valid Method to Study Neural Inter-brain Underpinnings of Social Interaction," *Frontiers in Human Neuroscience* 14 (2020): 39.
16. Recep Kocak, "The Effects of Cooperative Learning on Psychological and Social Traits among Undergraduate Students," *Social Behavior and Personality* 36, no. 6 (2008): 771–82.
17. Kocak.
18. Carlos Astete, Cristina Resino, Alina Boteanu, Maria Cañamero, "Application of a Cooperative Learning Approach for Training Residents in the Emergency Department and Comparison with Traditional Approach," *Emergencias* 27, no. 4: 231-235.
19. Rezvan Khoshlessan, "Is There a Relationship between the Usage of Active and Collaborative Learning Techniques and International Students' Study Anxiety?" *International Research and Review* 3, no. 1 (2013): 55–80.

20. Katrina J. Moffat, Alex McConnachie, Sue Ross, and Jillian M. Morrison, "First Year Medical Student Stress and Coping in a Problem-Based Learning Medical Curriculum," *Medical Education* 38, no. 5 (2004): 482–91.
21. Martyn Pickersgill, "Epigenetics, Education, and the Plastic Body: Changing Concepts and New Engagements," *Research in Education* 107, no. 1 (2020): 72–83.
22. Daniel Frías-Lasserre, Cristian Villagra, and Carlos Guerrero-Bosagna, "Stress in the Educational System as a Potential Source of Epigenetic Influences on Children's Development and Behavior," *Frontiers in Behavioral Neuroscience* 13, no. 12 (2018): 143.
23. Pickersgill.
24. Pickersgill.
25. Darlene Ciuffetelli Parker and Hillary Brown, eds., *Foundational Methods: Understanding Teaching and Learning* (Boston: Pearson, 2012).
26. Mina Tsay and Miranda Brady, "A Case Study of Cooperative Learning and Communication Pedagogy: Does Working in Teams Make a Difference?," *Journal of the Scholarship of Teaching and Learning* (2010): 78–89.
27. Sanikan Wattanawongwan, S. D. Smith, and Kimberly J. Vannest, "Cooperative Learning Strategies for Building Relationship Skills in Students with Emotional and Behavioral Disorders," *Beyond Behavior* 30, no. 1 (2021): 32–40.
28. David W. Johnson, Roger T. Johnson, and Edythe Johnson Holubec, *The Nuts and Bolts of Cooperative Learning* (Edina, MN: Interaction Book Company, 1994).
29. Sarentha Chetty, Varsha Bangalee, and Petra Brysiewicz, "Interprofessional Collaborative Learning in the Workplace:

A Qualitative Study at a Non-governmental Organisation in Durban, South Africa," *BMC Medical Education* 20, no. 1 (2020): 1–12.

CHAPTER 8: GETTING SMARTER AS A SOCIETY

1. Arline T. Geronimus, "The Weathering Hypothesis and the Health of African-American Women and Infants: Evidence and Speculations," *Ethnicity & Disease* (1992): 207–21.
2. Geronimus.
3. Gene Demby, "Making the Case That Discrimination Is Bad for Your Health," NPR, January 4, 2018, https://www.npr.org/sections/codeswitch/2018/01/14/577664626/making-the-case-that-discrimination-is-bad-for-your-health.
4. Sarah K. Letang, Shayne S-H. Lin, Patricia A. Parmelee, and Ian M. McDonough, "Ethnoracial Disparities in Cognition Are Associated with Multiple Socioeconomic Status-Stress Pathways," *Cognitive Research: Principles and Implications* 6, no. 1 (2021): 1–17.
5. Letang et al.
6. Elissa S. Epel, Elizabeth H. Blackburn, Jue Lin, Firdaus S. Dhabhar, Nancy E. Adler, Jason D. Morrow, and Richard M. Cawthon, "Accelerated Telomere Shortening in Response to Life Stress," *Proceedings of the National Academy of Sciences* 101, no. 49 (2004): 17312–15.
7. G. Tyler Lefevor, Caroline C. Boyd-Rogers, Brianna M. Sprague, and Rebecca A. Janis, "Health Disparities between Genderqueer, Transgender, and Cisgender Individuals: An Extension of Minority Stress Theory," *Journal of Counseling Psychology* 66, no. 4 (2019): 385.
8. Lefevor et al.
9. Lefevor et al.

10. Ian M. McDonough, "Beta-amyloid and Cortical Thickness Reveal Racial Disparities in Preclinical Alzheimer's Disease," *NeuroImage: Clinical* 16 (2017): 659–67.
11. Juliette McClendon, Katharine Chang, Michael J. Boudreaux, Thomas F. Oltmanns, and Ryan Bogdan, "Black-White Racial Health Disparities in Inflammation and Physical Health: Cumulative Stress, Social Isolation, and Health Behaviors," *Psychoneuroendocrinology* 131 (2021): 105251.
12. Letang et al., "Ethnoracial Disparities."
13. Letang et al.
14. Toni Schmader and Michael Johns, "Converging Evidence That Stereotype Threat Reduces Working Memory Capacity," *Journal of Personality and Social Psychology* 85, no. 3 (2003): 440.
15. Letang et al., "Ethnoracial Disparities."
16. Johnna R. Swartz, Ahmad R. Hariri, and Douglas E. Williamson, "An Epigenetic Mechanism Links Socioeconomic Status to Changes in Depression-Related Brain Function in High-Risk Adolescents," *Molecular Psychiatry* 22, no. 2 (2017): 209–14.
17. Edwin N. Aroke, Paule V. Joseph, Abhrarup Roy, Demario S. Overstreet, Trygve O. Tollefsbol, David E. Vance, and Burel R. Goodin, "Could Epigenetics Help Explain Racial Disparities in Chronic Pain?" *Journal of Pain Research* 12 (2019): 701.
18. Laura R. Cortes, Carla D. Cisternas, and Nancy G. Forger, "Does Gender Leave an Epigenetic Imprint on the Brain?" *Frontiers in Neuroscience* (2019): 173.
19. J. Guintivano, P. F. Sullivan, A. M. Stuebe, T. Penders, J. Thorp, D. R. Rubinow, and S. Meltzer-Brody, "Adverse Life Events, Psychiatric History, and Biological Predictors of Postpartum Depression in an Ethnically Diverse Sample of

Postpartum Women," *Psychological Medicine* 48, no. 7 (2018): 1190–1200.

CHAPTER 9: SEEING VALUE IN US ALL

1. Mitchell L. Yell, *The Law and Special Education* (Old Tappan, NJ: Merrill/Prentice-Hall, 1998).
2. Yell.
3. J. F. L., "Judge Lifts California Ban on IQ Tests for Black Children," *Pediatrics* 93, no. 1 (1994): 31, https://doi.org/10.1542/peds.93.1.31.
4. U.S. Department of Education, Office of Special Education and Rehabilitative Services, *Thirty-Five Years of Progress in Educating Children with Disabilities through IDEA*, 2010.
5. David J. Connor and Beth A. Ferri, "Integration and Inclusion—A Troubling Nexus: Race, Disability, and Special Education," *Journal of African American History* 90, nos. 1–2 (2005): 107–27.
6. Connor and Ferri.
7. National Center for Learning Disabilities, "Significant Disproportionality in Special Education: Current Trends and Actions for Impact," 2020.
8. National Center for Learning Disabilities.
9. Kelley Durkin, Mark W. Lipsey, Dale C. Farran, and Sarah E. Wiesen, "Effects of a Statewide Pre-kindergarten Program on Children's Achievement and Behavior through Sixth Grade," *Developmental Psychology* (2022).
10. Durkin et al.
11. Annette Lareau, "Unequal Childhoods," in *Unequal Childhoods* (Berkeley: University of California Press, 2011).
12. Fabiola Cineas, "The Future of New York City's Segregated 'Gifted' Programs," Vox, 2022.

CONCLUSION

1. Howard E. Gardner, *Frames of Mind: The Theory of Multiple Intelligences* (New York: Basic Books, 2011).
2. Daniel Goleman, *Emotional Intelligence: Why It Can Matter More than IQ* (London: Bloomsbury, 1996).
3. André Beauducel, Burkhard Brocke, and Detlev Liepmann, "Perspectives on Fluid and Crystallized Intelligence: Facets for Verbal, Numerical, and Figural Intelligence," *Personality and Individual Differences* 30, no. 6 (2001): 977–94.
4. Daniel Kahneman, *Thinking, Fast and Slow* (New York: Macmillan, 2011).

ABOUT THE AUTHOR

DR. RINA BLISS is associate professor of sociology at Rutgers University. Her research explores the personal and societal significance of emerging genetic sciences. Rina has written two books: *Race Decoded: The Genomic Fight for Social Justice,* revealing how genomics became today's new science of race; and *Social by Nature: The Promise and Peril of Sociogenomics,* which traces convergences in social and genetic science, and their implications for healthcare, education, criminal justice, and policymaking. She lives in Princeton, New Jersey, with her husband and three children.